Pathological Realities

forms of living

Stefanos Geroulanos and Todd Meyers, *series editors*

Pathological Realities

Essays on Disease, Experiments, and History

Mirko D. Grmek

*Edited, translated, and with an
Introduction by Pierre-Olivier Méthot
Foreword by Hans-Jörg Rheinberger*

FORDHAM UNIVERSITY PRESS
NEW YORK 2019

Frontispiece: Portrait of Mirko Grmek (1988). Fonds Mirko
Grmek/Archives IMEC; © John Foley/Opale/Leemage.

Fordham University Press has no responsibility for the persistence or accuracy of URLs for external or third-party Internet
websites referred to in this publication and does not guarantee
that any content on such websites is, or will remain, accurate or
appropriate.

Fordham University Press also publishes its books in a variety of
electronic formats. Some content that appears in print may not be
available in electronic books.

Visit us online at www.fordhampress.com.

Library of Congress Cataloging-in-Publication Data

Names: Grmek, Mirko D. (Mirko Dražen), 1924–2000, author. | Méthot,
    Pierre-Olivier, editor translator.
Title: Pathological realities : essays on disease, experiments, and history /
    Mirko D. Grmek ; edited, translated, and with an introduction by
    Pierre-Olivier Méthot ; foreword by Hans-Jörg Rheinberger.
Description: First edition. | New York : Fordham University Press, 2018. |
    Includes bibliographical references and index.
Identifiers: LCCN 2018011244 | ISBN 9780823280346 (cloth : alk. paper) | ISBN
    9780823280353 (pbk. : alk. paper)
Subjects: LCSH: Diseases—History. | Medicine—History. | Medical
    sciences—History. | Medicine—Historiography. | Medicine, Greek and
    Roman—History.
Classification: LCC R133 .G76 2018 | DDC 610.938—dc23
LC record available at https://lccn.loc.gov/2018011244

Printed in the United States of America
21 20 19 5 4 3 2 1
First edition

CONTENTS

Part IV. Memoricide: War and the Eradication of Cultural Memory

Hans-Jörg Rheinberger (Berlin)

I met Mirko Grmek twice in my life in person. Both encounters happened at turning points in my own scientific career. The first was at the Sixth Course of the International School of the History of Biological Sciences on the island of Ischia in 1988. At that time, I was still working as a molecular biologist in Berlin. The school was remarkably relaxed and dominated by Alistair Crombie's presence. It was also here that I made the acquaintance of Timothy Lenoir, who later invited me to spend a sabbatical at Stanford's Program in History of Science in 1989/1990. At the end of which I decided to leave the laboratory and to move into the history of science. Mirko Grmek was the gray eminence of the school, always keeping somewhat in the background but having an eye on everything.[1] Together with Bernardino Fantini, he had founded the school as well as the journal *History and Philosophy of the Life Sciences*, both housed at the Stazione Zoologica in Naples, in 1979, and he continued to run them throughout the 1980s with the help of Jean Ann Gilder and Christiane Groeben. The journal as well as the school are flourishing to this day and testify to Grmek's remarkable scientific entrepreneurship.

The second time I met Mirko was about a decade later, a few years before his untimely death, at the Fondation Louis-Jeantet, Geneva, during the Theory and Method in the Life Sciences workshop in 1996.[2] Now, I was preparing to move to the newly founded Max Planck Institute for the History of Science in Berlin and had just finished writing my book on experimental systems and their role for an epistemology of biology.[3] My impression of Mirko was again that of a grand seigneur, accompanied this time by French writer Louise Lambrichs.

The most lasting impression of Mirko Grmek's scholarship on me, however, came from a third, intellectual encounter. It concerns his publications of and on the laboratory notes of nineteenth-century French physiologist

Claude Bernard. Soon after I had started my case study on the history of in vitro protein synthesis research, I hit on Grmek's commented edition of Claude Bernard's *Cahier de notes, 1850–1860*, the so-called *Cahier rouge*.[4] He had edited and published it with Gallimard in 1965, while inventorying Bernard's papers left at the Collège de France shortly after he had settled in France. Carefully introduced and commented, these notes greatly encouraged me in my efforts to engage in laboratory microhistory of molecular biology. As Grmek emphasizes on the occasion of Bernard's work on the glycogenic function of the liver: "In this case, as probably in the historical analysis of all other scientific discoveries, it is of invaluable help to resort to original first-hand documents. The importance of systematic conservation of this kind of document, especially of laboratory journals, can hardly be overestimated."[5]

It is probably not overstated to consider Mirko Grmek as the founding father of historical laboratory microstudies, starting with his own book *Experimental Reasoning and Toxicological Research by Claude Bernard*, followed by Frederic Holmes's study, *Claude Bernard and Animal Chemistry*, and his concept of investigative pathway.[6] It was Grmek who had set Holmes on a track that proved extremely fruitful over the ensuing decades and that soon broadened into what came to be known as the "practical turn" in science studies in general.[7] As far as the relation between history of science and philosophy of science is concerned, Grmek followed the legacy of his elder Dutch colleague Eduard J. Dijksterhuis who once claimed that it was history that had to be considered itself as a sort of "laboratory of epistemology."[8]

Besides that, Grmek, himself a medical doctor by training, opened a new chapter in the history of medicine. Instead of focusing on illnesses as a human condition and the historically changing means of their conceptualization and remediation, he paved the way for looking at diseases as historical agents, both in their interactions with each other and their impacts on the course of human culture, be it plague in the Middle Ages or AIDS more recently. To cope with this challenge, he coined the notion of "pathocenosis," a concept formed in parallel to the ecological concept of biocenosis. Karl August Möbius created the latter term toward the end of the nineteenth century in order to characterize particular communities of organisms both in their interdependency and their dynamics.[9] Under the heading of pathocenosis, Grmek pursued the idea that diseases are a decisive factor and an integral part of what might be called an "ecohistory" of humankind. He traced that history in numerous books and interventions from antiquity to the present, and he mobilized not only written documents for that purpose but also documents of archaeology: the remnants of material culture.

May this collection of papers—addressing the history of science as well as medical and political history—help to keep the broad and innovative scientific legacy of Mirko Grmek inspiring and alive. He was a paragon of crossing boundaries: national, cultural, linguistic, as well as disciplinary borders.

Pathological Realities brings together Mirko Grmek's acclaimed research on Claude Bernard, AIDS, and the historiography of science and introduces some important contributions to the history of disease that remain largely unknown outside the French-speaking academic world. Assembled here for the first time, these essays were written and published over a period of some forty years in a variety of places, including ex-Yugoslavia, France, and the United States. Taken together, they range from the global history of disease concepts and the epistemology of early modern to nineteenth-century life sciences to the historiography of science and medicine and the Balkan War. Despite their inescapable differences in style, methods, and focus, these essays capture some of the main threads in Grmek's critical engagement with fundamental historical, philosophical, and scientific questions and their wider social and political implications. In addition, *Pathological Realities* provides a biographical overview of Grmek's life and work, although nothing short of a full biography could do justice to the scope of his contributions as a historian and a philosopher of the medical and biological sciences and to his complex personal trajectory.

In translating and editing this new collection of essays, I have tried to remain as close as possible to Grmek's original prose. Style and punctuation, however, had to be adjusted in some places for grammatical reasons and to clarify the meaning of the author as well as to ensure consistency throughout the book. In texts already in English, typos have been corrected and, given that many of the original texts included dates of birth/death for main authors, I provided those that were missing in the other essays. However, despite genuine efforts to obtain those, some are still missing. A few explanatory notes and references have also been added sparingly to the translated texts. As with other books in this series, notes in square brackets are my own.

Contrary to other French authors, Grmek's use of technical terms is sparse. Still, to increase readability I have rendered the French term *patho-cénose* as "pathocenosis" instead of "pathocoenosis," in contrast with earlier English translations of Grmek's work. The term *memoricide* remains the same in French and in English. The French *gnoséologique*, in turn, was rendered as "epistemological" or "theory of knowledge," depending on the context (see Chapter 7).

Although Grmek generally cites sources with accuracy, some references were sometimes incomplete or altogether missing in the original essays. Chapter 6, for instance, was devoid of bibliographical references and Chapter 7 included only a list of sources at the end. Drawing on the original published or unpublished French versions of these essays, I have included the missing references in the text. When published, Chapters 5 and 6 lacked section titles. Those included here are thus my own. The original, handwritten French version of Chapter 7 was found too late in the Fonds Grmek in Caen to allow for a whole new English translation. Instead, I updated the text by inserting a few missing paragraphs and sentences, by adding the relevant references, and by reformulating certain passages.

When publishing in English, Grmek often translated non-English sources himself. Whenever a modern and reliable English translation was available, it was used instead of Grmek's own translation—for instance, in Descartes's *Discourse on the Method* or Claude Bernard's *Introduction to the Study of Experimental Medicine*. The present edition also includes updated citations that were only partially translated and bibliographical references to works published in English since Grmek's original publication. Latin citations are provided with an English translation in notes or in parentheses.

Part of Chapter 1 appeared in translation in Mirko D. Grmek, *Diseases in the Ancient Greek World* (Baltimore, Md.: Johns Hopkins University Press, 1989). With the exception of the analytic definition of the concept of pathocenosis, however, the present book proposes a new and complete translation of Grmek's essay. The sections on ecology and immunology, omitted from the 1989 translation, are now available for the first time in English. Chapters 3, 4, 5, 6, and 7 of this book originally appeared in English in different journals and books. Chapter 3 was published in the *Journal of the History of Medicine and Allied Sciences* 50 (1995) and Chapter 4 appeared in the *Journal of the History of Biology* 1 (1968). The following three chapters were published in monographs or edited collections: Chapter 5 appeared in Mirko D. Grmek, *On Ageing and Old Age: Basic Problems and Historic Aspects of Gerontology and Geriatrics* (The

Hague: W. Junk, 1958); Chapter 6 was published in A. D. Breck and W. Yourgrau, eds., *Biology, History, and Natural Philosophy* (New York: Plenum, 1972); and Chapter 7 was included in Mirko D. Grmek, Robert S. Cohen, and Guido Cimino, eds., *On Scientific Discovery: The Erice Lectures 1977* (Boston: Reidel, 1981). To my knowledge, Chapters 2, 8, and 9 have never appeared in English translation.

I am grateful to the following publishers for granting the rights to republish and/or to translate the nine essays included here: Éditions de l'EHESS (Chapter 1), *History and Philosophy of the Life Sciences* (Chapter 2), Oxford University Press (Chapter 3), Springer (Chapters 4, 5, 6, and 7), Le Seuil (Chapter 8), and *Alliage: Culture-Science-Technique* (Chapter 9).

To some extent, this reader reflects my personal research interests (disease ecology, retrospective diagnosis, emerging infections, experimental practices in the life sciences, historiographical debates, etc.) through which I was initially drawn to Grmek's work. In selecting, translating, introducing, and editing *Pathological Realities*, however, I have sought to provide a nuanced, balanced, but also engaging picture of Grmek's scholarly contributions and research methods. I hope the resulting book would have pleased him and will be useful to students and scholars working at the interface of the human and natural sciences. At any rate, exploring Grmek's wide-ranging contributions was challenging and exciting, and I feel privileged to offer them a second life here today.

Mirko Grmek's Investigative Pathway

Pierre-Olivier Méthot

I

"Diseases only exist in the world of ideas." They are "explanatory models" of the world, not "constitutive elements of it."[1] When advancing these claims, Croatian-born historian of life sciences Mirko Dražen Grmek (1924–2000) was not denying the reality of individual illnesses but was referring to the "cultural fabrication" of disease concepts.[2] As he put it: "One can say that there exists a person who is sickly, coughs, spits blood, and grows thin, and one can say that the bacillus that pervades his organism and produces characteristic lesions on his lungs and other organs also exists in the strictest sense of the word, but one cannot say the same for tuberculosis."[3] As a separate entity, tuberculosis has no ontological existence; it belongs to a cultural and scientific network of bacteriological concepts and practices. Stating that diseases "do not flow directly from our experience" and emphasizing the incommensurable aspect of modern and ancient disease concepts would seem to deny the possibility of retrospective diagnosis.[4] However, drawing on literary works, scientific texts, and artistic and biological artifacts, Grmek's scholarship pieces together the "traces of the past in the present" and reconstructs the "pathological reality" of particular historical moments, places, and peoples.[5]

Educated between Zagreb and Paris in the medical, natural, and human sciences, Grmek's work places these realms into dialogue and goes beyond disciplinary and methodological borders. Straddling the sciences and the humanities and defying traditional historiographical classifications, his writings range from those of a biologist and a philologist to those of a philosopher and a historian—and are sometimes all of them at once. At the cultural and geographical level, Grmek also crossed several borders, often depicting himself as a "citizen of the world."[6] Being fluent in at least eight European languages, in addition to ancient Greek and Latin, allowed him to engage directly with primary sources and made him a privileged interlocutor to scholars in science and the humanities across Europe and North America for nearly half a century. Like other prominent medical historians of the earlier generation—such as Owsei Temkin (1902–2002), Georges Canguilhem (1904–1995), Erwin Ackerknecht (1906–1988), or George Rosen (1910–1977)—Grmek trained as a physician before turning to history and had a detailed knowledge of the technical aspects of illness. But in contrast to most of these figures, he was also well-versed in ancient medicine, a world authority on French physiologist Claude Bernard (1813–1878), a scholar on seventeenth-century life sciences, a leading researcher of AIDS and emerging infections, not to mention a commentator on the collapse of Yugoslavia.[7]

Grmek's commanding stature as a historian rests, in addition, on his exceptional capacity for work and scholarly production, as a list of more than twenty-five monographs spanning more or less the whole history of the life sciences testifies (Appendix A). As an editor, his work includes twenty-five more volumes (Appendix B) and two encyclopedias, in Croatian and in French (Appendix C).[8] Yet even a compound list of authored and edited volumes only partially illustrates the scope of his research interests, whose bibliography includes over one thousand items published in Croatian, French, English, German, Italian, Greek, Russian, and Portuguese.[9] Among historians of medicine and science, Grmek also stands apart because he engaged with contemporary social and political issues, such as war, genocides, and global pandemics. The writing of history is, for him, an exercise in critical thought in the present. Grmek, indeed, was long persuaded that the history (of science) is a "militant discipline,"[10] that is, a vehicle for "a critical appreciation of modern problems," as he put it in his introductory lecture at Berkeley in 1967, and not a matter of passive "melancholic contemplation."[11] Through detailed historical case studies, his scholarship reveals the interconnections of diseases, societies, and medical theories. "Transformations in our knowledge about health and disease," he writes, "are at the same time transformations of power."[12]

In the present day, historians of medicine and biology remember him especially for writing the first history of AIDS, for his study of Bernard's experimental methodology, and for his contributions to the field Stephen Boyden (b. 1925) has termed "biohistory"—although these are just a few facets of his work.[13] Drawing upon the concept of "pathocenosis," a construct of his own, Grmek saw biological relationships and diseases as crucial factors in "forging human destiny." In contrast to traditional history, which focuses on the fate of "great men," he put forth a global historical vision in which diseases interact with each other and are to be understood as "mass phenomena" with an impact at the social, economic, scientific, and even political levels.[14] In a sense, Grmek's work anticipates later, more deterministic narratives linking human history with biological events such as William McNeill's *Plagues and People* or Alfred Crosby's *The Columbian Exchange*.[15]

The singular trajectory that took Grmek from Yugoslavia and Italy to the academic culture of postwar France placed him at the crossroads of different intellectual trends and made him an influential figure in the history of science and medicine during the second half of the twentieth century. Combining many theoretical approaches and bringing up new research questions, Grmek carefully avoided committing himself to a single method: "To favor an approach, and within it a particular theory, to make it the exclusive means of investigation," he asks, "isn't this a sin of excessive epistemological optimism?"[16] Venturing onto untrodden historical (and historiographical) paths, he created what his younger American colleague, the historian of science and medicine Frederic L. ("Larry") Holmes (1932–2003), called an "investigative pathway," a term referring to the general investigative movements in which individual scientists take part.[17] But in spite of, or perhaps because of, his several achievements, scholars have rarely attempted to articulate Grmek's distinctive vision of the history of science and medicine with all its tensions, contradictions, and ambiguities.[18] And although a number of his books and articles have appeared in English translation over the years, some major strands of his published work are still unavailable, dispersed, or simply unknown. Uniting some of those strands in a single place, this collection of essays that spans five decades builds on scholarship already available in English and adds to it, offering a broader picture of Grmek, which covers deep epistemological changes in historical conceptions of disease as well as major developments within cultures of the life sciences and their historiography.

Opening with "Preliminaries for a Historical Study of Diseases" (1969),

"The Concept of Emerging Disease" (1993), and "Some Unorthodox Views and a Selection Hypothesis on the Origin of the AIDS Viruses" (1995), this first section introduces Grmek's notions of pathocenosis and emerging infections, which are illustrated with both historical and contemporary cases. The second section of the collection turns to experimental approaches in biology and medicine. "First Steps in Claude Bernard's Discovery of the Glycogenic Function of the Liver" (1968), "The Causes and the Nature of Ageing" (1958), and "A Survey of the Mechanical Interpretations of Life from the Greek Atomists to the Followers of Descartes" (1972) examine turning points in the history of biological thought. The third section introduces Grmek's perspective on the epistemology and methodology of history of science and is composed of a single but densely argued chapter titled "A Plea for Freeing the History of Scientific Discoveries from Myth" (1981). Finally, "A Memoricide" (1991) and "Dubrovnik: The Slavic Athens" (1995) form the last section of the book and consider Grmek's intellectual engagement with wars, cultural memory, and the politics of knowledge. Presenting a general overview of his life that long remained "shrouded in an aura of mystery," this introductory essay outlines the contours of Grmek's investigative pathway in situating his most significant contributions within some of the main historiographical trends in history and epistemology of science and medicine of the past century and introduces the nine essays included in this collection.[19]

II

EARLY LIFE, WORLD WAR II, AND MEDICAL STUDIES

Mirko Dražen Grmek was born into a bourgeois family (his father, Milan Grmek [1892–1956], was a lawyer) in Krapina, Kingdom of Serbs, Croats, and Slovenes (territory of today's Croatia), on January 9, 1924, before moving to Zagreb a few years later.[20] In addition to being fluent in Croatian and German, the young Grmek also learned basic French by the age of seven thanks to his mother, Vera Santovać (1902–1997), who had studied literature, theater, and arts and made him read French poetry. At secondary school, setting up a small chemistry laboratory in his bedroom, he also demonstrated a marked interest for the natural sciences.[21] Struck in his youth by a form of pulmonary tuberculosis, he was forced to stay at a sanatorium for a few months at the age of fifteen. Despite his alleged poor health, Grmek was drafted into the army of the Independent State of Croatia when the country entered the Second World War in 1941. Refusing to fight for the Axis alliance, and with the help of an army general who knew his father, he orchestrated his own transfer to the military academy of Turin (Regia Accademia di Genio), a nonfascist polytechnic institution temporarily relocated to Lucca, Tuscany, because of the war. In addition to the natural sciences and mathematics, Grmek also learned Italian and developed a deep affinity for the country.

Grmek remained a student at the Italian academy until the fall of Mussolini in July 1943. After that, he joined the Italian Resistance. Escaping with a few fellow students from Lucca, then surrounded by German troops, he was forced to travel through Italy before crossing over to Switzerland via an underground network, where he stayed in a number of work camps and refugee camps until the end of the war.[22] In 1944, Grmek was transferred to Küssnacht am Rigi, near Luzern, an idyllic region in the center of the country. The Swiss authorities recognized his rank of lieutenant in the Italian army and, although he was constantly guarded, he could freely read philosophy and science books, available through the YMCA services. Here, he began writing essays and poems, which, he said, helped him "overcome the atrocities of war."[23] This "blessed" period of his life, as he called it, led Grmek to part ways with both metaphysics and religion and left him convinced that "the goal of human life is to forget it has none."[24] Once the war was over, he left Switzerland and, traveling through France, reached recently liberated Paris, where he worked for some time as an interpreter for an American military commission before returning to Yugoslavia via Marseilles and Bari, Italy.

By the end of the war Grmek had obtained the rank of lieutenant in the French Liberation Army.[25] Upon his return to Zagreb, he married his first wife, Sida Vallić, with whom he had a son, and entered the Zagreb Medical School (1946–1951).[26] Drawn to endocrinology, an exciting field at the time, he also worked in psychiatry in Zadar and Zagreb hospitals.[27] Although medical practice appealed to him because it seemed like an "authentic source of information about human nature," the hierarchical aspect of the medical system in place soon reminded Grmek of the army and failed to conform to his ideal.[28] Furthermore, echoing the earlier words of his long-distance mentor, the Swiss-born medical historian Henry E. Sigerist (1891–1957),[29] Grmek noted that a career in medicine requires specialized knowledge in one particular field, whereas he longed for more general or "synthetic knowledge."[30] Like Sigerist, Grmek's rather exceptional linguistic skills permitted him to pursue his project of developing a broad knowledge of medical history—

from pre-Hippocratic medicine to contemporary biomedical sciences—and not to remain confined to a particular historical period or problem.

As a medical student, he published his first article in 1946, at the age of twenty-two, and his ongoing participation in conferences across Eastern and Western European countries led him to earn a reputation among professional historians of science and medicine. His desire to study Croatian medical history from the Middle Ages and Renaissance[31] was supported by some prominent Yugoslav scholars, including his "master" Ladislav Lujo Thaller (1891–1949), the founder of the history of medicine in the Kingdom of Serbs, Croats, and Slovenes, Italian medical historian Arturo Castiglioni (1874–1953), Dalmatian medical historian and priest Miho Barada (1889–1957),[32] and also by Sigerist, who expected "a great deal from him."[33] In a historical survey on the teaching of medical history in Yugoslavia, Grmek praises Thaller's leadership in the creation of a Section for the History of Medicine in the Medical Association of Croatia, which led to his chairing of the Ninth International Congress of the History of Medicine in Zagreb, Belgrade, Sarajevo, and Dubrovnik in 1938.[34]

Grmek's contributions on the work of Dubrovnik scholars Santorio Santorio (1561–1636) and Giorgio Baglivi (1668–1707) were also noticed by Andrija Štampar (1888–1958), then dean of the School of Public Health in Zagreb and president of the Croatian Academy of Science.[35] When at the Ministry of Public Health in the 1920s, Štampar advocated social hygiene measures and eugenicist interventions (such as premarital compulsory examination) based on an organicist view of society.[36] As an international figure and a pioneer of social medicine, Štampar had been involved in the League of Nations and played a critical role in the creation of the World Health Organization (WHO). While serving as its first secretary, he was influential in drafting the definition of *health* written in the preamble of the WHO Declaration (1948).[37]

After graduating with an MD in 1951, Grmek completed a compulsory one-year internship in Zadar Hospital and, thanks to the city's very rich library,[38] he discovered that Napoleon created a medical faculty in 1811, a finding he would draw upon in his PhD dissertation.[39] At the end of his internship the next year, Thaller helped Grmek join the Croatian Medical Association. Following this, he progressively abandoned clinical practice to focus on the history of medicine, which he saw as a bridge between the humanities and the natural sciences.[40] Also in 1952, in a letter supporting Grmek's application to the Institute for the History of Medicine at Johns Hopkins—then directed by American medical historian Richard Shryock (1893–1972)—Sigerist reported that Štampar was "very anxious to have a chair of the history of medicine created in Zagreb" and thought Grmek was the "best man for the post," though he was still in need of "specialized training."[41] Shryock was interested in having a "Yugoslav scholar resident in the Institute," but Štampar's project of securing funds to send Grmek to Baltimore was unsuccessful because, that year, scholarships available were allocated to American and Canadian graduate students only.[42] In a quick turn of events, however, and with the support of Sigerist, Štampar helped Grmek to create an Institute for the History of Medicine in Zagreb and to become its first director in 1953.

One year after writing his habilitation dissertation "on the life and works on the medieval medical writer Dominiko of Dubrovnik (1558–1613),"[43] Grmek was elected as editor-in-chief of the main Croatian medical journal, *Liječnički vjesnik* (The medical courier), and Štampar entrusted him with teaching his Introduction to Medicine course. Grmek's course, however, was not in the history of medicine—then taught by neuropsychiatrist Lavoslav Glesinger (1901–1986). Indeed, according to the teaching program in the Zagreb Medical School, the goal of Introduction to Medicine was to introduce first-year students to fundamental "medical ideas," to develop "a correct attitude towards the tasks and social function of the doctor," and to understand "the position of the doctor in relation to the development of a proper system of public health."[44] Reflecting the physical division of labor between history and medical deontology broadly construed, the former was taught at the Zagreb Medical Faculty, whereas the latter took place at the School of Public Health.

Drawing on archival sources found in Zadar and in Paris, Grmek completed his doctoral dissertation—"Medical Faculties in Dalmatia in the Period of French Administration (1806–1813)"—in 1958, becoming the first scholar in Yugoslavia to earn a PhD in the history of science. Two years later, his administrative responsibilities increased as he took the directorship of the newly created Institute of the History of the Natural, Mathematical, and Medical Sciences and became full professor of medical history in Zagreb. "From historian of medicine," he commented later, "I became more and more historian of science."[45] Within the biological and other natural sciences, Grmek was always particularly interested in the history of ageing, sexuality, and death.[46] Perhaps as a result of his earlier interest in endocrinology, he investigated the leading scientific hypotheses concerning the nature and causes of ageing, which culminated in a book,[47] also in 1958, where he defines *ageing* as the "progressive and irreversible changing of the structures and functions of the living system."[48]

The death of his father and Štampar, the deteriorating relationship with his wife, his meeting with biologist Danièle Guinot in 1958, as well as a desire to explore new, untapped historical sources more freely, eventually led Grmek to pursue his career outside the geopolitical frontiers of Yugoslavia.[49] Following a two-month research fellowship from the Centre national de la recherche scientifique (CNRS) to work at the Bibliothèque nationale de France (Paris) in the summer of 1960 under the mentorship of Annalist historian Fernand Braudel (1902–1985), Grmek traveled regularly to the French capital.[50] In 1963, the Collège de France invited him to organize and catalog the unpublished manuscripts and notes of Claude Bernard, a challenge he longed for and that turned him into a world expert on the life and works of the French physiologist.[51] Prior to his departure from Yugoslavia, and paralleling his growing interest for "epistemology and the analysis of concepts," Grmek established friendships with Pierre Huard (1901–1983), a scholar on Southeast Asian history of medicine, and Jean Théodoridès (1926–1999), a historian of medical parasitology.[52] Retrospectively, Grmek attributed his transition toward the study of the history of biological sciences in France to this earlier period where he progressively moved from the history of sanitary conditions and social medicine to the "historical vicissitudes of general biological ideas."[53]

FROM ZAGREB TO PARIS AND BERKELEY:
TOWARD THE HISTORY OF BIOLOGICAL CONCEPTS

Throughout the 1960s and 1970s, Grmek's scholarship expanded from ancient Slavic medicine, medical deontology, and social medicine to the history of surgery, seventeenth- and nineteenth-century physiology, disease concepts, gerontology, pathography, cell theory, Chinese sphygmology (the examination of the pulse), medical geography, disease ecology, epistemology, as well as the development of quantitative approaches in the biological sciences. Some of his most original contributions during this period concern the history of disease and Claude Bernard's experimental practices.[54] Two of these essays, reprinted here, stand out as they offer key illustrations of Grmek's investigative and historiographical methods. Drawing on nascent ecological theories, history of civilizations, and parasitology,[55] Grmek called for a "natural history of disease"[56] and coined the concept of "pathocenosis" to capture an idea that, he later commented, "appeared vaguely and was escaping us."[57] Written in a Braudelian style with subtle structuralist undertones, Grmek's article focuses on the "slowly evolving endemic diseases" in their capacity to "affect the biological potential of a society to a much deeper and more durable extent."[58] In the essay on laboratory practices, in contrast, he uses the unpublished papers, laboratory notebooks, and manuscript notes of the French physiologist Claude Bernard to challenge prevailing historical narratives based on great discoveries and great men. "How interesting it is to measure the extent to which a great scientist reconstructs his own previous thoughts to fit his later point of view," he writes.[59]

Developing an international network of collaborators, Grmek's work and influence soon went beyond the borders of France and other European countries and led him to actively conduct several stays in the United States from the mid-1960s to the early 1970s.[60] Grmek first completed a six-month residency at the University of California–Berkeley as visiting associate professor in 1967, upon the invitation of historian of science Roger Hahn (1932–2011). His introductory lecture laid out three main theoretical orientations in the field of the history of science: technical, conceptual, and sociological—written by scientists, philosophers, and trained historians of science, respectively. These viewpoints, he stressed, "are not really in conflict," foreshadowing his later criticism of the "epistemological optimism" of adopting a unique means of investigation, and he emphasized "the legitimacy of concurrent attitudes" while following "the general patterns of Alexandre Koyré and Canguilhem."[61] During this period, Grmek also presented his work on Claude Bernard at Berkeley's History of Science Dinner Club founded by George Sarton (1884–1956).[62] In September the same year, he flew from Paris to Detroit to participate in a workshop on the historical and epistemological aspects of the biological sciences. Hosted by philosopher, historian, and physicist Wolfgang Yourgrau (1908–1979), the meeting was also attended by, among others, biochemist and Nobel Prize laureate Albert Szent-Györgyi (1893–1986),[63] system theorist Ludwig von Bertalanffy (1901–1972), and paleontologist George G. Simpson (1902–1984).[64] Grmek's paper, reprinted here, traced the main ideas underpinning mechanical interpretations of life in the history of biology. During this trip to the United States, Grmek traveled to New York, where he was invited by population geneticist Theodosius Dobzhansky (1900–1975) and evolutionary biologist Francisco Ayala (b. 1934) to give a talk at the Rockefeller Institute. Shortly afterward, he took part in the meeting of the American Academy of Psychoanalysis with physician and medical historian Iago Galdston (1895–1989).[65]

In February 1968, historian of science Charles D. O'Malley (1907–1970) extended an invitation to Grmek to participate in an international symposium on the history of medical education at the University of California at

Irvine. Grmek's talk was on the history of medical education in Russia. This meeting brought together many of Grmek's colleagues and mentors from Europe, such as Luigi Belloni (1914–1989), Francisco Guerra (1916–2011), Pierre Huard, and F. N. L. Poynter (1908–1979).[66] Before returning to Europe, Grmek stopped at Princeton University upon the invitation of Thomas Kuhn (1922–1996) to deliver a lecture on Bernard.[67] In April 1969, as visiting lecturer at the Harvard Medical School, Grmek worked at the Countway Library on the nature and causes of ageing. He shared his insights into these processes in a conference titled "The Search for a General Theory of Ageing" that remained unpublished.[68] Though harmful for the individual, he argues, ageing and death are nonetheless "useful and necessary" at the level of the species as they ensure its continuation and a greater "biological adaptability."[69] Reflecting on how this process enables life to "escape ageing," Grmek considered ageing at the molecular and cellular levels in terms of "loss of genetic information," and explained sexuality as a "rejuvenation process that is fundamental to the maintenance of organismic negentropy."[70] During his stay at Harvard, Grmek also attended the congress of the American Association for the History of Medicine at Johns Hopkins, which gave him the opportunity to visit the Institute for the History of Medicine, then directed by Temkin, where he once wanted to receive advanced training.

One year after giving one of the opening addresses during the inauguration of the Morris Fishbein Centre for the History of Science and Medicine in Chicago, historian of science Seymour Chapin (1927–1995) helped Grmek to become visiting professor at the California State College in the spring of 1971.[71] This time, his lectures considered the social and political consequences of natural sciences such as physics and chemistry, with a particular focus on the discovery of radioactivity and its applications. "Science," he claims, "is neither good nor bad, but its application can have perverse effects . . . and only science can provide the means to fight back its misuses."[72] A highlight of this professorship in Los Angeles was the invitation by molecular biologist Max Delbrück (1906–1981)—who attended one of Grmek's public lectures—to give a paper on "Descartes as a Biophysicist" in his seminar at Caltech, Pasadena, in May 1971.[73] In his autobiographical writings, Grmek acknowledged this opportunity that acquainted him with the laboratory "where modern molecular biology was born."[74] Pursuing his collaboration with his American colleagues later in the mid-1970s, Grmek worked with historian of science Robert S. Cohen (1923–2017) following an advanced course on the history and epistemology of scientific discovery in Erice, Sicily, which led to a book in the Boston Studies in the Philosophy of Science series (of which a chapter is reprinted here).[75] This synopsis of Grmek's activities reveals that his reputation in North America before the 1980s was neither modest nor centered on philology; at the time, Grmek was mostly known within and outside Europe as a scholar in the history and epistemology of the biological sciences.[76]

PROFESSIONALIZING THE HISTORY OF SCIENCE

Grmek's arrival in Paris coincides with the professionalization of the history of science and its departure from general history and philosophy of science, as well as from the history of medicine.[77] The two main research centers in history of science in the French capital were the Institut d'histoire des sciences—created in 1932 by Abel Rey (1873–1940) and directed by Canguilhem during Grmek's residency, which had a clear philosophical orientation—and the Centre de recherche d'histoire des sciences et des techniques,[78] founded in 1958 by Braudel and Alexandre Koyré (1892–1964) and close to the Centre international de synthèse, which brought together scholars with a more scientific orientation like René Taton (1915–2004) or Pierre Costabel (1912–1989) and literary scholars turned historians of science, such as Jacques Roger (1920–1990). As a protégé of Koyré, Braudel, Taton, and Canguilhem—and a colleague of Roger and Costabel—Grmek navigated between these two poles often perceived as being in tension[79] and drew much inspiration from the resources of these rich intellectual environments, where he actively contributed in terms of teaching and research.[80]

Like Thaller, Štampar, and Sigerist, these scholars exerted an influence on Grmek's own view of history of science, although he did more than follow the concepts and methods put forth by his Parisian mentors. For instance, while acknowledging his intellectual debt to Braudel, his several studies of Bernard's experiments contributed to "micro-history" and did not belong to the "*longue durée*" paradigm.[81] Also, in spite of his insistence, like Koyré, on "scientific revolutions,"[82] Grmek recognized "major historical continuities"[83] and considered that revolutionary moments and continuity always coexist in history of science and medicine.[84] And even as he spoke of Canguilhem as the "indisputable master of his generation of historians and philosophers of science,"[85] Grmek's attitude toward "the Cang" and his school was hardly that of a diligent pupil.[86] Presenting history of science as the "laboratory of epistemology," he sought to free it from pernicious "methodological myths"[87] and the normative frameworks of philosophers of science.[88] "The history of scientific concepts" of the Paris school, he told his students in Berkeley, "is not the only correct way of coming to an understanding of the history of science."[89] Like Roger, finally, Grmek thought that an "entirely non-presentist history" was impossible because "historians cannot entirely forget who they are, nor what they know."[90] But in contrast to Roger, Grmek also considered that current medical research provides "invaluable help to the historian of diseases" and "that any medical discovery can be helpful to historical projects."[91]

Grmek defended his thesis, "Recherches expérimentales et toxicologiques chez Claude Bernard" (Experimental reasoning and toxicological research by Claude Bernard), written under Canguilhem's guidance in 1971, obtaining his second PhD in the history of science.[92] Linking Bernard's psychology to his philosophy of science, Grmek's historico-epistemological work insisted on what he calls the "lived experience of discovery" (*le véçu de la découverte*) and on the roles of imagination, creativity, and even error in paving the way to scientific discoveries. This lived experience, he argues, is essential to the discovery process, which "cannot be the sort of formal logic" concerned "exclusively with matters of justification or validation."[93] The role of justification in science, he explains, is twofold: in the lived experience of his work "the researcher wishes to convince himself," while in his published work "he wishes to convince others."[94] Uncovering the underlying image of the scientist-as-artist that informed Bernard's day-to-day practice, Grmek's research on laboratory notebooks and case history inspired a new historiographical approach to study science "in-the-making" that was successfully applied by Larry Holmes and by Gerald Geison (1943–2001) on the likes of Antoine Lavoisier (1743–1794), Hans A. Krebs (1900–1981), and, of course, Louis Pasteur (1822–1895).[95]

Succeeding Huard in 1973, Grmek became director of studies at the École Pratique des Hautes Études (EPHE), where his "Thursday seminars" attracted a wide array of students coming from classics, history, and philosophy.[96] Interested in promoting international collaborations, he also occupied several administrative posts in scholarly societies and devoted considerable time to editorial activities.[97] In addition to the publication of an encyclopedia on the history of science and techniques in French, one of his main achievements was the foundation (with Bernardino Fantini) of the international journal *History and Philosophy of the Life Sciences* in 1979. This thriving, multilingual journal, branching off from a scientific periodical in marine biology edited in Naples, became a reference point for historical-critical reflections on biology for European and North American scholars.[98] From the late 1970s to the early 1990s, he extended his teaching to an even broader audience by directing the Ischia Summer School on the History of the Biological Sciences.[99] Located in the former home of German marine biologist Anton Dohrn (1840–1909), the founder of the Stazione Zoologica in Naples, it served to foster the career of many scholars.[100]

Pleading for the autonomy of the history of science,[101] Grmek often defended its "professional specificity" by arguing that only the "historical method" can provide objectivity in the study of sources,[102] though he insisted that objectivity in historical research remains an ideal.[103] Resenting the "malaise" between philosophy and history of science in the 1970s, Grmek believed that the two disciplines could still occupy common ground (for instance, with regard to the origins of scientific hypotheses)[104] but, since epistemology is often out of touch with the "historical reality," he remained doubtful about its "heuristic value" for history of science.[105] Responding to Foucault's argument in "Cuvier's Situation in the History of Biology," for instance, he criticized the French philosopher for proposing a "logical construction" devoid of actual "historical content."[106] In a 1990 interview, Grmek claimed that Foucault's work "is lacking from a methodological point of view," that it is "based on personal intuitions, brilliant ideas that are immediately generalized; well," he continued, "this approach is fundamentally anti-scientific."[107] Sensitive to the "erudite" approach pioneered by Charles Daremberg (1817–1872)[108] and Braudel's "scientific history," Grmek positioned the history of science as a field closer to scientific work than to philosophy—a marked contrast with Canguilhem and the French school of historical epistemology for whom "The object in the history of science has nothing in common with the object of the science."[109] In his "Prolégomènes à une histoire générale des sciences" (Prolegomena for a history of science), Grmek even states that the growing place of general historians and philosophers of science might put the nascent field of history of science at risk.[110]

Grmek's particular approach to history of science and medicine also stood in an uneasy relation with trends in the social history of medicine that developed from the early 1970s to the late 1980s. Roy Porter's influential "history of the patient," for instance, stands in marked tension with Grmek's project of retrospectively diagnosing diseases in the past and a history of medical concepts.[111] Grmek is not concerned with exploring the world (or the "voices") of the patient or inclined to practice "history from below." Furthermore, Grmek's scholarship only rarely attends to the fields of psychology or psychiatry (as Porter but also Foucault did) and is mostly concerned with the rise and advances of medical sciences, which, for Grmek, parallel the laboratory tradition leading to, and inherited from, Claude Bernard. Grmek belongs to a generation of scholars where medical (or scientific) training and acquaintance with a body of technical knowledge was essential to the rightful practice of medical history.[112] Perhaps surprisingly, however, because of his emphasis on biopathological mechanisms and his rejection of the "social construction" metaphor, the attempt of social historian Charles Rosenberg (b. 1941) to "frame" diseases in history may be more directly compatible with Grmek's view that disease conceptualizations are influenced by the underlying "physical and biological reality."[113] Be that as it may, Grmek resisted the various trends in the social history of medicine during the 1970s and 1980s and remained both faithful to "conceptual" history and fearful of a "history of medicine without medicine."[114] In a letter to his former graduate student Jacalyn Duffin (b. 1950),[115] for example, he advises her not to be too impressed by the "present exaggeration of the sociological approach in the history of medicine" and urges her to continue working "in the spirit of a 'technical,' truly 'medical,' and 'conceptual' history."[116] In Grmek's medical universe, such "conceptual" or "medical" history was "neither anecdotic nor purely social, as could be the history of hospital institution," but was foremost oriented toward "guiding ideas."[117]

FROM ANCIENT DISEASES TO EMERGING INFECTIONS

Bringing his scientific and medical expertise to bear on philological and historical problems, Grmek aspired to contribute to the study of the "pathological reality" of the prehistorical, archaic, and classical Greek world, which led to a long and fruitful collaboration with Jacques Jouanna (b. 1935), Fernand Robert (1908–1992), Danielle Jacquart (b. 1947), Simon Byl (b. 1940), Philippe Mudry (b. 1939), and other scholars in ancient medicine working in Paris. Aware of the epistemological issues inherent to the practice of retrospective diagnosis, Grmek placed high hopes on the development of paleopathology[118] in order "to reconstruct the nosological reality of a society that is twenty-five centuries apart from ours."[119] The study of bones, mummies, and DNA fragments, he thought, could be used to "skirt the pitfalls" of written sources, to go beyond medical discourses, and to provide a some-what more "objective" approach to the study of ancient diseases.[120] Like other paleopathologists, Grmek's use of modern knowledge to diagnose diseases in the past stems from his conviction that the human body has not signifi-cantly changed in two thousand years and that "pathological reactions . . . are the same now as in the past."[121] This assumption, in turn, compelled him to look for changes in diseases and disease patterns in the complex relations between organisms, germs, and the environment.[122] Drawing on the concept of pathocenosis, he attempted to bring epidemiologists, geographers, and his-torians closer together in elucidating changes and continuity in the disease landscape over time.[123]

Bringing together over one decade of scholarship in philology and medical history, in addition to summarizing recent findings in microbiology, paleo-pathology, genetics, anthropology, epidemiology, and history of ideas, Grmek shed new light on the "pathological reality" of the ancient Greek Mediter-ranean world in his 1983 magnum opus *Les maladies à l'aube de la civilisation occidentale* (*Diseases in the Ancient Greek World*). The reception of the book in France was positive, especially among philologists and philosophers. For ex-ample, Robert hailed Grmek's new book as an "event in Greek studies," while Canguilhem praised the "astounding richness of sources" and the "mastery in their exploitation."[124] Published by Johns Hopkins University Press six years later, this book aligned Grmek's work with the scholarly tradition of medi-cal history at Hopkins established by Sigerist and carried over by Temkin and Ludwig Edelstein (1902–1965), but it did not attract the attention of historians of biological sciences even though it drew upon the most recent advances in biology. The originality and significance of the book, articulating geographical, biological, and social factors, was perhaps less obvious to con-temporary English-speaking scholars who struggled with the term *pathoceno-sis*, on the one hand, and were by then already familiar with "bio" perspectives on human history, thanks to the work of McNeill, Crosby, and Boyden and the rise of paleopathology, on the other.[125] As Colin Jones rightly points out, the concept of pathocenosis, at least in the English-speaking world, "seems never to have achieved the kind of breakthrough for which Grmek must have hoped."[126] More important, historians have been quick to take his assumption about the "ecological stability" of diseases in Greek antiquity to task.[127]

The book on the history of the Greek pathocenosis was followed six years later by a biohistorical study (translated into English and published by Princeton University Press in 1990) on the first "postmodern" disease: AIDS. Drawing again on pathocenosis but also on contemporary virology and molecular biology, the author argued—with "mordant Gallic wit,"[128] to use Porter's words—that the pandemic originated because of the progressive elimination of competing pathogens, rather than of the sudden evolution of more virulent germs. "AIDS," Grmek comments in an interview, "is the price we pay for having radically perturbed millenary ecological equilibria."[129] Thinking of diseases as an interconnected system of ecological relations, he claims that "whenever a disease disappears, another eventually makes its appearance."[130] In a critical review of the book, Gerald Oppenheimer (1922–2016) grants that it provides a fascinating account of the origin of the pandemic but notes that Grmek's hypothesis that HIV "persisted in silent parasitism" was "speculative," even though it provides an initial explanation of the pandemic.[131] Other historians in France and in the United States, like Michaël Pollak (1948–1992) and Thomas Laqueur (b. 1945), however, found Grmek's rhetoric objectionable.[132] But even Laqueur notes that though Grmek's analysis could easily be dismissed as "old-fashioned homophobia unsuccessfully masquerading as value-free medicine," his more general point was that humans have always upset existing ecological equilibria, which, in turn, creates new patterns for disease transmission.[133] In the end, "stripped from its normative baggage," Laqueur writes, "Grmek is making an important argument about how new patterns of human behavior create new pathways for pathogens."[134] Grmek's approach certainly deserves to be reconsidered at the present with all the transformations in the experimental, therapeutic, population, and geopolitical aspects of the disease.

EMERGING INFECTIONS, ETHNIC CLEANSING, AND WAR AS A SOCIAL DISEASE

When officially retiring from the École Pratique in 1989, Grmek was honored with a Festschrift edited by his long-time friend and successor at EPHE, the historian and classicist Danielle Gourevitch (b. 1941)[135] and continued to lecture across Europe and North America, to write on Bernard's experimental method, and to study the representations of disease in artwork ("icono-diagnostic").[136] In 1991, he was awarded the Sarton Medal by the History of Science Society, the highest prize in the field. In his Sarton address, Roger Hahn singled out Grmek's "steadfast devotion to the classical canons of history" to explain why "he stands as such an admirable model for our profession."[137] Two years later, at the invitation of Larry Holmes, he delivered the

1993 Henry E. Sigerist lecture in the History and Sociology of Medicine at Yale University on the origins of the AIDS viruses (reprinted in this volume). During the 1990s, Grmek edited a three-volume encyclopedia published in Italian, French, and (partly) English, covering the entire course of medical history, and created a new collection in the history of medicine titled *Penser la médecine* at Fayard (Paris). The figure of Grmek as being primarily a medical historian crystalized, in France and abroad, during this last period of his life, although the major part of his career has been in the history of biology and in the history of science and techniques.

In close dialogue with American virologist Stephen Morse (b. 1951) from Columbia University (who recommended the English translation of the book on AIDS),[138] Grmek was among the few French scholars who, during the late 1980s and early 1990s, called attention to the issue of "emerging" infectious diseases such as Legionnaires' disease, Ebola, and Lassa fever.[139] Connecting the process of emergence to changes in host-parasite relationships, technological advances, novel conceptualizations of pathological phenomena, and changes in pathocenosis, Grmek argued that "new diseases" are (usually) not the result of newly evolved virulent organisms but the result of microbial germs that gained entry into a new susceptible host population, often thanks to human activities.[140] In 1992, he brought together prominent historians (including William McNeill), epidemiologists, and virologists to place emerging infections in historical, critical, and scientific perspective during an international meeting in Annecy at the Charles Mérieux Foundation. By foreseeing the possibility of new diseases flourishing following the development of cultural practices and ecological changes, Grmek could be rightly depicted as a forerunner of the modern concept of emerging disease. However, in a posthumous publication he gave credit to Morse and bacterial geneticist Joshua Lederberg (1925–2008) with "the first successes in the epidemiological use of the term 'emergent'" and considered that the conference on emerging viruses they organized in Washington, D.C., in 1989 "was particularly significant and in many respects even premonitory."[141]

In his last interview in late February 2000, a few weeks before his death, Grmek discusses with journalist Mirko Galić (b. 1945) how Franjo Tuđjman (1922–1999)—a World War Two partisan, a general in the Yugoslav army, and, later, the founder of the Croatian Democratic Union, a center-right political party—offered him the post of ambassador of Croatia in France.[142] Declining Tuđjman's offer, Grmek became involved in the International Croatian Initiative in 1992, along with a number of Croatian scholars in exile.[143] Engaging intellectually with the dissolution of Yugoslavia with French writer and literary scholar Louise Lambrichs (whom he married in 1999), he depicts the Balkan War as a "social disease,"[144] denounces the ideology of "ethnic cleansing,"[145] and coins the concept of "memoricide" (reprinted here) to decry the purposeful eradication of cultural memory.[146] Adopting a resolutely scientific standpoint on the conflict, Grmek views the war in former Yugoslavia as a "social pathology," a "psychosocial epidemic," and even a "moral disease of human groups" that can be studied using scientific methods such as empirical psychology.[147] Resorting to the analogy between cells and citizens and their relations to the whole, he argues that when "some influential and important people are ill, displaying characteristics of psychosocial and social pathology, the whole society can suffer, as when some diseased cells cause a reaction to the overall organism."[148] Looking at larger ensembles as models for judgments of health and disease, Grmek asks: "Why not . . . speak of diseases of society, in the literal sense?"[149] In looking at war as a social pathology, Grmek's political writings embrace an organicist view of society such as those that flourished in the late nineteenth and early twentieth centuries—including in the works of his former master in Zagreb, Andrija Štampar.[150]

Despite the steady flow of publications that characterized his writings in this period, the 1990s were difficult for Grmek because of the war in his home country. The last two years of his life were especially painful. His mother, who lived with him in Paris since the mid-1970s, passed away in 1997. And after losing his only son, who died of a heart failure in 1998, he was diagnosed with a form of amyotrophic lateral sclerosis one year later, an illness that progressively prevented him from moving and carrying out his work. Owing to the fast-declining state of his bodily functions, he penned several notes and letters about his illness to his friends and colleagues, who often wrote back affectionately,[151] and published "The Third Scientific Revolution," a reflexive essay on history of science and often seen as his intellectual testimony.[152] Refusing to be permanently placed on a respirator, he died in Paris on March 6, 2000, and he is buried in Montparnasse Cemetery.[153]

III

The first section of chapters straddles both the local and the global study of disease. In the opening essay, Grmek introduces the term *pathocenosis*, by which he means the interaction of different diseases within a given population over a specific period of time. Initially presented during a symposium on medicine and culture in London organized by the Wellcome Library and the Wenner-Gren Foundation for Anthropological Research in September 1966 and chaired by Poynter and Galdston, the concept appeared in "Préliminaires d'une étude historique des maladies," which was published in *Les Annales* as part of a series on material life and biology edited by Braudel.[154] Basing this concept on contemporary ecological theories, medical geography, as well as parasitology, Grmek argues that no disease should be studied or understood in isolation but only as part of a larger community of illnesses that impinge on each other in such a way as to revert over time to a dynamically stable condition, despite occasional crises. While "analytical" approaches treat epidemics and other disease events in isolation from each other, "synthetic" approaches—as promoted by Grmek—study how they bear on each other. With the concept of pathocenosis, Grmek provides historians of medicine, philosophers, sociologists, and historians of ideas with a powerful intellectual tool to study dynamically the historical specificity of past and present diseases.

The concept of pathocenosis was introduced at a historical moment dominated by an understanding that infectious diseases had largely been conquered. The second chapter, written some thirty years later, reformulates and applies the idea that diseases form complex interconnected systems in the context of the global threat of emerging infections. Here, Grmek argues that the ancient notion of "new disease" should be replaced with the modern term "emerging disease," as suggested by Morse and Lederberg.[155] "Le concept de maladie émergente" also introduces a framework to clarify the multiple ways in which a disease is "emerging" and illustrates Grmek's engagement with contemporary medical and social problems. In contrast with the classic epidemiological definition of emergence based on an increase in incidence (i.e., the number of new cases), Grmek broadens the term *emergence*, which, in his world shaped by biological relations, can refer to the transfer of infectious germs from animal to human populations (or from one human population to another), to changes in the clinical manifestations of a disease, to the evolution of new pathogenic agents, and to the crafting of a new nosological concept. Thus, emergence might be described as a process that is as biological as it is intellectual.

Building on *History of AIDS*, the third essay deals with the origin and emergence of pathogenic viral strains (HIV1–HIV2) prior to the onset of the world pandemic and critically considers the main hypotheses regarding the origin of those emerging strains from the combined lens of epidemiology, evolutionary biology, and medical history.[156] "Some Unorthodox Views and a Selection Hypothesis on the Origin of the AIDS Viruses" traces the conceptual, political, and technological difficulties in conceptualizing AIDS as a disease category and argues that medical progress, by facilitating the transmission of infections through blood transfusion and suppressing competing infectious parasites, has led to a rupture in twentieth-century "pathocenotic equilibrium." Invoking both sociological and biological factors, which are "inextricably intertwined" and related to "postwar civilization," Grmek emphasizes "the biological links" between infectious diseases, changes in human behavior, and technological progress (e.g., blood transfusion) in explaining the emergence of the pandemic. The provocative (and sometimes offensive) rhetoric deployed by Grmek in this essay should not detract readers from its main point: The spreading of infectious germs is only a precondition for the onset of an epidemic, which results from a concatenation of biological and social factors.

The second section of this collection concentrates on the development of experimental practices and concepts in the life sciences. Drawing on laboratory notebooks and research protocols, "First Steps in Claude Bernard's Discovery of the Glycogenic Function of the Liver" follows Bernard's day-by-day reflections on physiological problems. Published in the first issue of the *Journal of the History of Biology* upon the invitation of Harvard historian of science Everett Mendelsohn (b. 1932), the chapter contrasts Bernard's unpublished manuscripts that suggest a "gradual development of his discoveries" and his published work that, in contrast, shows "a tendency toward a secondary rationalization," that is, "a very strong *post hoc* simplification of facts." Focusing on Bernard's research spanning a few (decisive) weeks in August 1848, Grmek traces the neglected complexities of the process by which the French physiologist came to view the presence of sugar in blood as neither pathological nor accidental but instead as normal physiological phenomenon.[157] Grmek's study based on his cataloging of Bernard's manuscript for the Collège de France was a pioneering attempt at using scientific notebooks for writing a new genre of histories of science that led to the "practice turn" in history and philosophy of science.

Chapter Five in this collection, "The Causes and the Nature of Ageing," consists in one of Grmek's long-standing research topics. After introducing a formal definition of ageing as the "progressive and irreversible changing of the structures and functions of living systems," this chapter focuses on the most significant scientific and philosophical attempts to grasp its core nature. Following the historical trajectory of two competing hypotheses that hark back to Hippocrates, Aristotle, and Galen, the chapter examines whether ageing results from a "gradual loss of vital energy" or from the progressive "accumulation of something deleterious for the organism" and concludes that any model of ageing must consider a variety of external and internal causal factors interacting together. Traced throughout the nineteenth and up to the mid-twentieth century, the examination of these models ends in the 1950s when Sir Peter Medawar (1915–1987) and George C. Williams (1926–2010) put forth evolutionary hypotheses of ageing.[158] Drawing on a vast array of multilingual primary and secondary sources, Grmek's wide-ranging intellectual history of biological ideas goes beyond Western scientific worldviews and gives credit to Soviet and Eastern European scientists for furthering our understanding of the nature and cause of ageing.

The third chapter of this section, "A Survey of the Mechanical Interpretations of Life from the Greek Atomists to the Followers of Descartes," was first presented by Grmek at the Academy of Science in Moscow in 1966. Here, the author provides a clear and informed historical overview of mechanical models of living phenomena from antiquity to the "first biological revolution" in the seventeenth and eighteenth centuries and explores how vitalistic biology has persistently challenged such mechanical models. Drawing on the writings of Canguilhem and insisting on the pivotal role of quantitative methods in the "revival of the new mechanics" in the seventeenth century, Grmek emphasizes the limitations of this paradigm and stresses, for example, that any mechanistic theory of life is a "methodological fiction." Descartes's concept of "beast-machine," he says, was precisely hard to accept for his contemporaries because it was an idea that could only be realized by "engineers of the future." In this essay, Grmek points out the cyclical recurrence of certain concepts in the history of biology, teases out the role of analogies and models in shaping biological knowledge, and showcases the enduring debate between mechanism and vitalism from the Greeks until the present.

The third section of the book outlines the epistemological and methodological principles that inform Grmek's work as a historian of science and medicine. First presented at the Ettore Majorana Centre in Erice, Italy, during the International School for the History of Science in 1977, "A Plea for Freeing the History of Scientific Discoveries from Myth" focuses on the growth of science, that is, on science understood as an ever "expanding body of knowledge."[159] Drawing on the genetic epistemology of Jean Piaget (1896–

1980), Grmek views "cognitive processes" and "somatic regulating events" as isomorphic and argues that "the acquisition of knowledge bears the essential characteristics of *biological growth*." Summarizing the main points of his work on Bernard and other scientific figures, and engaging with Thomas Kuhn's and Karl Popper's (1902–1994) works and the ideas of "paradigm shift" and epistemological "falsification," he exposes "mercilessly"[160] over a dozen of entrenched "methodological illusions," such as the myth of the "straight road to truth," the myth of "continuous evolution and permanent revolution," as well as the myth of a "strictly logical nature of scientific reasoning."

Looking at his later political writings, the fourth and final section of the volume concludes with two short essays written during the Balkan War in the 1990s. The first one, "Un mémoricide," initially published in December 1991 in the French political journal *Le Figaro*, appeared a few months after Dubrovnik was bombed. Grmek's piece introduces the concept of "memoricide" to French readership in order to characterize the purposeful eradication of cultural memory. For Grmek, this concept encompasses not only the destruction of written documents or cultural artifacts but also the systematic demolition of historical monuments and, indeed, of all traces of the past in the present. As conveyed by the etymology of the term, *memoricide* is the cultural equivalent of genocide. The other essay, "Dubrovnik: l'Athènes slave," appeared in 1995 in *Alliage*, a journal of critical analysis founded by French physicist and philosopher Jean-Marc Lévy-Leblond (b. 1940). In this ode to Dubrovnik, Grmek recounts how the ancient Republic of Ragusa historically worked as a "cultural relay" between East and West, allowing the free circulation (in both directions) of people, knowledge, instruments, medicines, and ideas. Recalling the long medical tradition of the city, he distills key scientific ideas of scholars such as Roger Joseph Boscovich (1711–1787) and Giorgio Baglivi and their relation to their country before turning to the "historical sense of the actual bombing." Drawing on the concept of memoricide to denounce the organized destruction of historical monuments, such as churches, libraries, and hospitals (including lazaretto), the essay ends with a plea for making Dubrovnik an "artistic center" again that would help the city reclaim its "vocation of being an interface between civilizations."

Appendix A (Books Published as Author, Including Translations in Other Languages)

1. *Santorio Santorio i Njegovi Aparati i Instrumenti* [The Istrian physician Santorio Santorio, his apparatuses and instruments]. Zagreb: Jugoslavenska akademija znanosti i umjetnosti, 1952.

2. *On Ageing and Old Age: Basic Problems and Historic Aspects of Gerontology and Geriatrics.* The Hague: W. Junk, 1958.

3. *Les sciences dans les manuscrits slaves orientaux du Moyen Age* [The sciences in East Slavic manuscripts during the Middle Ages]. Paris: Université de Paris, 1959.

4. *Uvod u medicinu* [An introduction to medicine]. Zagreb: Medicinska knjiga, 1961.

5. *Les reflets de la sphygmologie chinoise dans la médecine occidentale* [The reflection of Chinese sphygmology in Western medicine]. Paris: Specia, 1962.

6. *L'introduction de l'expérience quantitative dans les sciences biologiques* [The introduction of quantitative experimentation in biological sciences]. Paris: Université de Paris, 1962.

7. *Gerontologiya, utchenie o starosti i dolgoletti* [Gerontology: The science of ageing and longevity]. Moscow: Nauka, 1964.

8. *Les conditions sanitaires et la médecine en Dalmatie sous Napoléon 1er (1806–1813)* [Sanitary conditions and medicine in Dalmatia under Napoleon 1st (1806–1813)]. Paris: Specia, 1964.

9. with Pierre Huard, *Mille ans de chirurgie en Occident, V^e–XVe siècle* [One thousand years of surgery in the Occident, fifth–fifteenth centuries]. Paris, Dacosta, 1966.

10. with Pierre Huard, *La chirurgie moderne, ses débuts en Occident: XVIe, XVIIe, XVIIIe siècle* [Modern surgery: Its beginnings in the West: Sixteenth, seventeenth, eighteenth centuries]. Paris, Dacosta, 1968.

11. with Pierre Huard, *Sciences, médecine et pharmarcie de la Révolution à l'Empire (1789–1815)* [Science, medicine, and pharmacy from the Revolution to the Empire (1789–1815)]. Paris: Dacosta, 1970.

12. *Raisonnement expérimental et recherches toxicologiques chez Claude Bernard* [Experimental reasoning and toxicological research in Claude Bernard]. Geneva: Droz, 1973.

13. *Psicologia ed epistemologia della ricerca scientifica: Claude Bernard, le sue ricerche tossicologiche* [Psychology and epistemology of scientific

research: Toxicological research in Claude Bernard]. Milan: Episteme, 1976.

14. with Pierre Costabel, *L'Académie internationale d'histoire des sciences: Cinquante ans, 1927–1977* [The International Academy of the History of Science: Fifty years, 1927–1977]. Paris: Vrin, 1977.

15. *Les maladies à l'aube de la civilisation occidentale. Recherches sur la réalité pathologique dans le monde grec préhistorique, archaïque et classique* [Diseases at the dawn of Western civilization: Research on the pathological reality in the prehistoric, archaic, and classic Greek world]. Paris: Payot, 1983.

16. *Le malattie all'alba della civiltà occidentale. Ricerche sulla realtà patologica nel mondo greco presistorico, archaico e classico* [Diseases at the dawn of Western civilization: Research on the pathological reality in the prehistoric, archaic, and classic Greek world]. Bologna, Italy: Il Mulino, 1985.

17. *Nuove prospettive per la storia della malattie antiche* [New perspectives on the history of disease in antiquity]. Naples, Italy: Guida, 1988.

18. *AIDS. Storia di une epidemia attuale* [History of AIDS: Emergence and origin of a modern pandemic]. Rome: Laterza, 1989.

19. *Bolesti u osvit zapadne civilizacije. Istrazivanja patolške stvarnosti u grčkom prehistorijskom, arhajskom I klasičnom dobu* [Diseases at the dawn of Western civilization: Research on the pathological reality in the prehistoric, archaic, and classic Greek world]. Zagreb: Globus, 1989.

20. *Diseases in the Ancient Greek World*. Baltimore, Md.: Johns Hopkins University Press, 1989.

21. *Histoire du sida: Début et origine d'une pandémie actuelle* [History of AIDS: Emergence and origin of a modern pandemic]. Paris: Payot, 1989.

22. *History of AIDS: Emergence and Origin of a Modern Pandemic*. Princeton, N.J.: Princeton University Press, 1990.

23. *La première révolution biologique. Réflexions sur la physiologie et la médecine au XVIIe siècle* [The first biological revolution: Thoughts on physiology and medicine in the seventeenth century]. Paris: Payot, 1990.

24. *Claude Bernard et la méthode expérimentale* [Claude Bernard and the experimental method]. Paris: Payot, 1991.

25. *Povijest side. Počeci, š irenje i uzroci jedne epidemijske bolesti naš ih dana* [History from the side: The beginnings, the spread, and the causes of an epidemic disease in the present]. Zagreb: Globus, 1991.

26. *Prva biološ ka revolucija. Razmiš ljanja o fiziologiji i medicini 17. stoljeća* [The first biological revolution: Thoughts on physiology and medicine in the seventeenth century]. Zagreb: Globus, 1996.

27. with Antun Budak, *Uvod o medicinu* [Introduction to medicine]. Zagreb: Globus, 1996.

28. *Le chaudron de Médée. L'expérimentation sur le vivant dans l'Antiquité* [Medea's cauldron: Experimentation on the living in antiquity]. Le Plessis-Robinson, France: Synthélabo, 1997.

29. *Le legs de Claude Bernard* [The legacy of Claude Bernard]. Paris: Fayard, 1997.

30. *La vita, le malattie et la la storia* [Life, disease, and history]. Rome: Di Renzo, 1998.

31. with Danielle Gourevitch, *Les maladies dans l'art antique* [Diseases in antique art]. Paris: Fayard, 1998.

32. with Louise L. Lambrichs, *Les révoltés de Villefranche, Mutinerie d'un bataillon de Waffen-SS, september 1943* [The rebels of Villefranche. The mutiny of a battalion of Waffen-SS, September 1943]. Paris: Seuil, 1998.

33. *Mistika riječi. Lirski ciklusi* [The mystic of words: The lyrical cycles]. Zagreb: Ceres, 1999.

34. *La guerre comme maladie sociale et autres textes politiques* [War as a social disease and other political texts]. Edited by Louise L. Lambrichs. Paris: Seuil, 2001.

35. *La vie, les maladies et l'histoire* [Life, diseases, and history]. Edited by Louise L. Lambrichs. Paris: Seuil, 2001.

Appendix B (Books Published as Editor, Including Translations in Other Languages)

1. *Robert Koch: Die Aetiologie der Tuberkulose* [Robert Koch: The etiology of tuberculosis]. Zagreb: Medicinska knjiga, 1951.

2. with Stanko Dujmušic, *Iz Hrvatske medicinske prošlosti. Spomen-knjiga Zbora liječnika Hrvastske* [On Croatian medical past: Jubilee Collection of the Society of Physicians in Croatia]. Zagreb: Zbor liječnika Hrvatske, 1954.

3. *Iz Hrvatska medicinska bibliografija. Bibiographia medica Croatica* [Croatian medical bibliography]. Zagreb: Jugoslavenska Akademija, 1955.

4. with Pierre Huard, *Le premier manuscrit chirurgical turc rédigé par*

Charaf ed-Din (1465) et illustré de 140 miniatures [The first Turkish chirurgical manuscript, written by Charaf ed-Din (1465), illustrated with 140 miniatures]. Paris: Dacosta, 1960.

5. with Pierre Huard, *Dessins scientifiques et techniques de Léonard de Vinci* [Scientific and technical drawings of Leonardo da Vinci]. Paris: Dacosta, 1962.

6. *Rasprave i gradja za povijest nauka* [Studies and materials to serve the history of science], vol. 1. Zagreb: Yugoslav Academy of Science, 1963.

7. with Pierre Huard, *Charles Estienne: La dissection des parties du corps humain* [Charles Estienne: The dissection of human body parts]. Paris: Cercle du livre précieux, 1965.

8. *Claude Bernard: Cahier de notes 1850–1860, édition intégrale du "cahier rouge"* [Claude Bernard: Notebook 1850–1860, complete edition of the "red notebook"]. Paris: Gallimard, 1965.

9. *Claude Bernard: Notes, mémoires et leçons sur la glycogenèse animale et le diabète* [Claude Bernard: Notes, memoirs, and lessons on animal glycogenesis and diabetes]. Paris: Cercle du livre précieux, 1965.

10. *Notes inédites de Claude Bernard sur les propriétés physiologiques des poisons de flèches (curare, upas, strychnine et autres)* [Unpublished notes by Claude Bernard on the physiological properties of poisonous arrows (curare, upas, strychnine and others)]. Paris: Specia, 1966.

11. *Serving the Cause of Public Health: Selected Papers of Andrija Štampar.* Zagreb: School of Public Health, 1966.

12. *Catalogue des manuscrits de Claude Bernard* [Catalog of Claude Bernard's manuscripts]. Paris: Collège de France and Masson, 1967.

13. *Srednjovekovne rasprave Jakobela Vitturija Trogiranina o liječenju lovnih ptica I konja* [The medieval treatises on falconry and hippiatry of Jacobelus Vitturi from Trogir]. Zagreb: Veterinarski fakultet, 1969.

14. *Claude Bernard: Notes pour le Rapport sur les progrès de la physiologie* [Claude Bernard: Notes for the report on the progress in physiology]. Paris: Paillart, 1979.

15. *Hipppocratica. Actes du Colloque hippocratique de Paris (4–9 septembre 1978)* [Hippocratica: Proceedings of the Hippocratic Colloquium in Paris (September 4–9, 1978)]. Paris: CNRS, 1980.

16. *Radovi sa medjunarodnog Simpozija u povodu 600-te obljetnice dubrovačke karantene, Dubrovnik, 29. I 30. rujna 1977* [Proceedings of the international symposium on the occasion of the 600th anniversary of

the quarantine in Ragusa, Dubrovnik]. Zagreb: Yugoslav Academy of Arts and Science, 1980.

17. with Robert S. Cohen and Guido Cimino, *On Scientific Discovery: The Erice Lectures*. Dordrecht, Netherlands: Reidel, 1981.

18. with L. Boulle, C. Lupovici, and J. Samion-Contet, *Laennec: catalogue des manuscrits scientifiques* [Laennec: Catalog of scientific manuscripts]. Paris: Fondation Singer-Polignac and Masson, 1982.

19. with Guido Cimino and Vittorio Somenzi, *La scoperta scientifica. Aspetti logici, psicologici e sociali* [Scientific discovery: Logical, psychological, and social aspects]. Rome: Armando, 1984.

20. with M. Gjidara and N. Simac, *Etničko čišcenje. Povijesni dokumenti o jednoj srpskoj ideologiji* [Ethnic cleansing: Historical documents on a Serbian ideology]. Zagreb: Globus, 1993.

21. with M. Gjidara and N. Simac, *Le nettoyage ethnique. Documents historiques sur une idéologie serbe* [Ethnic cleansing: Historical documents on a Serbian ideology]. Paris: Fayard, 1993.

22. *Storia del pensiero medico occidentale. Antichità e Medioevo* [Western medical thought: From antiquity to the Middle Ages]. Rome: Laterza, 1993.

23. *Histoire de la pensée médicale en Occident. Antiquité et Moyen Age* [Western medical thought: From antiquity to the Middle Ages]. Paris: Seuil, 1995.

24. *Storia del pensiero medico occidentale. Dal Rinascimento all'inizio dell'Ottocento* [Western medical thought: From the Renaissance to the Enlightenment]. Rome: Laterza, 1996.

25. *Histoire de la pensée médicale en Occident. De la Renaissance aux Lumières* [Western medical thought: From the Renaissance to the Enlightenment]. Paris: Seuil, 1997.

26. *Storia del pensiero medico occidentale. Dall'età romantica alla medicina moderna* [Western medical thought: From Romanticism to modern science]. Rome: Laterza, 1998.

27. *Western Medical Thought from Antiquity to the Middle Ages*. Cambridge, Mass.: Harvard University Press, 1998.

28. *Histoire de la pensée médicale en Occident. Du romantisme à la science moderne* [Western medical thought: From Romanticism to modern science]. Paris: Seuil, 1999.

29. *Arétée de Cappadoce, Des causes et des signes des maladies aigües et chroniques* [Aretaeus the Cappadocian: On the causes and symptoms of acute and chronic diseases]. Geneva: Droz, 2000.

30. with Jacques Jouanna, *Hippocrate, Épidémies, Livres V et VII* [Hippocrates, epidemics, books V and VII]. Paris: Les Belles Lettres, 2000.

31. with Josip Balabanić, *O ribama I školjkašima dubrovačkog kraja / Korespondencija Jakov Sorkočević-Ulisse Aldrovandi (1580–1584)* [About the fish and shellfish of the Dubrovnik area / Correspondence between Jakov Sorkočević and Ulisse Aldrovandi (1580–1584)]. Zagreb: Dom isvijet, 2000.

Appendix C (Scientific Editor of Encyclopedias)

1. *Medicinska enciklopedija* [Medical encyclopedia], 10 vols., 1957–1965; 2nd ed., 6 vols. 1967–1970. Zagreb: Leksikografski zavod.

2. *Encyclopédie internationale des sciences et des techniques* [International encyclopedia of sciences and techniques], 12 vols. Paris: Presses de la Cité, 1969–1973.

PART ONE

Pathocenosis: Diseases in History

ONE

Preliminaries for a Historical Study of Diseases

In order to understand the real forces that forge human destiny, we must not neglect a major biological factor: disease. At the individual level, the importance of pathology is a banal fact. The dawn of the study of the incidence and impact of illness on personal lives corresponds to the first steps in medicine, as well as to the early stages of art and philosophical thinking. By attempting to understand the influence of diseases on the activities of celebrities, early writers thought that it might be possible to attain "great history."[1] Could the illnesses of the leaders determine the fate of the people? Greek historiographers were convinced of this, a legitimate thought in a period when historical predominance was attributed to a few leaders of personal genius. What would have been the fate of the classical world if Alexander the Great had not abused alcoholic beverages and suffered an early death from violent fever? And was the fall of the Roman Empire hastened, or even provoked, by epilepsy and other morbid afflictions of the Julio-Claudian dynasty?[2] Those questions are naïve, like the joke about Cleopatra's nose.[3] Yet serious historians are still debating the political consequences of Napoleon I's gastric troubles, of Napoleon III's urolithiasis, of Henry VIII's syphilis, and of Lenin's final treacherous illness.[4]

The influence of illness on the individual is not necessarily an evil. It can mark a positive turn in one's personal history and affect the collectivity by creating a condition, or at least a stimulus, for creative art, scientific investigation, or even social activities (reformer, religious, etc.). Whence springs the interest in studies, which today are called *pathography*.[5]

Traditional history, which describes the most striking military, cultural, and political events and pays tribute to the "great men," stands in contrast to a more complete vision of humanity's past, a history of daily life, namely a sociological approach mainly centered on the fate of the "average man." In light of this historical synthesis, diseases play a fundamental role as mass phenomena affecting the economy, demographic movements, and social mores.[6]

There are, of course, epidemics whose amplitude and devastating consequences are so great that their historical role seems obvious to all observers. Notably, the best descriptions of ancient epidemics of exceptional importance do not come from professional physicians but from historians or other scholars. Suffice it to remember the description of the "Plague" of Athens by Thucydides and Boccaccio's and Ibn Battuta's accounts concerning the "Black Death." Military historians cannot ignore the extent to which past military strategies depended on typhus, nor to what extent colonial expansion was hampered by yellow fever and other tropical diseases.[7]

If the influence of these disasters is clear, the same clarity does not apply to the slowly evolving endemic diseases that affect the biological potential of a society to a much deeper and more durable extent. The writer observing from within a group suffering from a chronic disease can evaluate only with difficulty the extent and consequences of the pathological factor. Concerning European populations, the appreciation of the social importance of diverse, chronic diseases could be made only after the fact—that is, through historical thought.

Until the present day, the historical study of diseases was almost exclusively *analytical*; that is to say that the development in time of each disease was studied individually. Already used by Gruner in *Morborum Antiquitates*,[8] this approach was later followed by Hirsch,[9] Creighton,[10] Corradi,[11] Sticker,[12] Rolleston,[13] Laignel-Lavastine and his colleagues,[14] Pazzini,[15] and many others. Even more recent authors such as Bett,[16] Ackerknecht,[17] or Henschen[18] rarely leave the beaten path.

While it has undeniable advantages, this analytical method is nevertheless flawed as it leaves in the dark facts concerning the reciprocal influences of several diseases. True, some specialists in medical geography, led by Finke[19] and guided by his rules regarding the composition of what was once called "medical topography," could not help noticing the interconnectedness of diseases.[20] This tendency toward the integration of all the morbid phenomena in a given time and place has pervaded recent studies on disease ecology (Jacques M. May[21] and others), but historians seem not to have drawn its full lesson.

In the historical study of diseases, the analytical approach is certainly the simplest and most fruitful. A synthetic approach cannot claim to replace it but can only complete it. To make this new approach more accessible, certain concepts medical historians have not yet used must first be defined. I hereby formulate the three following propositions:

1. *Pathocenosis* is the ensemble of pathological states present in a specific population at a given moment in space and time;
2. The frequency and overall distribution of each disease depends, in addition to endogenous and ecological factors, on the frequency and the distribution of all the other diseases;
3. A pathocenosis tends toward a state of equilibrium that is especially perceptible under stable ecological conditions.

It is fascinating and instructive to study, on the one hand, the synchronic relationship of diseases within a specific historical timeframe and, on the other, to follow the diachronic changes of these ensembles. But if the term *pathocenosis* includes the notion of synchronic relations, a new term is needed to designate its diachronic changes, that is, to provide a temporal dimension to the concept of pathocenosis. To this end, I propose the expression: *dynamics of pathocenosis*.

Two diseases belonging to the same pathocenosis can be in a state of symbiosis, antagonism, or indifference to each other—which paradoxically appears to be the most common and the most difficult relationship to demonstrate.

Cases of symbiosis are numerous and well known. Usually they result from an etiological connection at the social level: the same environmental conditions favor two or more diseases—for instance, the diseases resulting from stress in a technologically advanced society. It can also be a genetic link; for example, several hereditary syndromes are known to have numerous and separable nosological components. More important are the cases where sensitivity toward one illness is acquired at the individual level simply because of previous suffering from another. The examples are too numerous to spend time enumerating them. The "second" illness can be so frequent that it is considered a delayed specific manifestation of the first illness. Mumps orchitis or diphtheritic myocarditis or endocarditis following an attack of acute articular rheumatism fall in this category. Of course, it can be a unique disease from the pathogenetic point of view, but it does not appear as such in statistics on morbidity. Finally, symbiosis can also result from a more complex situation: for example, avitaminosis, anemias, and typhus sustain each other on both the social and the individual levels.

Historians will certainly be more fascinated by the antagonism between diseases. Who has not heard of Wagner-Jauregg's therapeutic procedure based on the antagonism between acute fevers (notably malarial) and tertiary syphilis?[22] But does this antagonism exist on the level of pathocenosis, and if so, why not invoke this fact to explain, even partially, the difference between the treponematosises of the Old and New Worlds? Physicians in the nineteenth century believed in the antagonism between malaria and tuberculosis; phthisics were sent for cures to malarial areas, but outcomes were scarcely brilliant. However, what is of interest to historians here is absolutely not the results of individual therapeutics. Our interest is turned toward the possibility of explaining a few major epidemiological events. For example, let us ask a particularly tricky and double question: Why did leprosy and plague disappear from Western civilization long before adequate treatments were developed?

To explain the "spontaneous" end of an epidemic today, preference is given to two types of hypotheses. First, this end could be due to the creation of a state of equilibrium between the host and the pathological agent. The majority of a population acquires either protection (after a violent phase, the infection becomes latent; it is without clinical consequences while preventing reinfection; let us call this hypothesis no. 1a) or immunological resistance, which diminishes the clinical reaction while at the same time blocking proliferation of the parasitic micro-organism (hypothesis no. 1b). It is easy to "understand" the biological mechanism of the phenomena by invoking the neo-Lamarckian schema (that is, thanks to the intrinsic adaptive properties of living beings) or by referring to the currently more popular neo-Darwinian schema (that is, the selection operated at the level of the molecular structures of the genetic code). Sabin's experiments show that it is possible to obtain, through selection, a strain of mice that resists infection by a dose of yellow fever virus that was absolutely fatal for controls.[23] A long time ago, the illustrious microbiologist Theobald Smith (1859–1934) expressed the opinion, shared by most of his colleagues, that there exists in the relationship between the host and the parasite a natural tendency toward mutual tolerance. It is in the "biological interest" of the microbe not to remain too virulent, because killing its host is often a form of suicide.[24] As with Sabin's mice, yellow fever can be so inoffensive for the inhabitants of certain African countries that it is exceedingly difficult to make the diagnosis: for these populations, it has become an "inapparent" illness.[25] The second group of hypotheses postulates an antagonistic activity between diverse micro-organisms—in short, in the phenomena known under the name of antibiosis (hypothesis no. 2). Hypotheses 1 and 2 are not necessarily mutually exclusive, since it is possible that the infection by one microbe protects or conveys resistance against another.

It seems probable that hypothesis 1a explains the historical evolution of several plagues. René Dubos (1901–1982) showed it well with the example of tuberculosis,[26] of certain gastrointestinal infections, of toxoplasmosis, and of a few other epidemiological entities.[27] To a certain extent, this hypothesis can be applied to all attempts to understand the historical fluctuations of plague and leprosy during the medieval endemic period; however, it cannot explain the almost total disappearance of these two diseases from the majority of European territory. If only two factors are at play, biological equilibrium will never be established by the total annihilation of one of them. The intervention of a third party is required to obtain this result. Concerning plague, this complementary factor could be one of the vectors known to inhabit the very complex epidemiological cycle of this disease. It would seem that for leprosy no such intermediaries exist. Nevertheless, modern researchers are fairly attracted by the idea of the intervention of a third organism. Therefore, today we are on the threshold of an unexpected solution to the problem of the disappearance of leprosy in Western Europe.

Lowe and McNulty,[28] Floch,[29] Chaussinand,[30] and others have shown that the primary infection by tuberculosis can render Mitsuda's reaction positive to lepromin and that it is most probably followed by a relative immunity to leprosy; BCG vaccination clearly diminishes the risk of an open infection by Hansen's bacillus.[31] It therefore seems possible, even probable, that the retreat of leprosy in the Occident was linked to the rise of tuberculosis, which coincided with the social, economic, and demographic transformations of the fifteenth century. Sigerist was correct in believing that the disappearance of leprosy was due to the ravages of another disease, but he was mistaken when he blamed plague, which would have provoked massive mortality in the leprosariums.[32] The eviction of leprosy took place in a more subtle and complicated manner—that is, through competition between two related mycobacteria, a competition whose outcome depended on numerous ecological factors and on the dynamics of the pathocenosis of Europe in its entirety.

I have deliberately raised the problem of the simultaneous extinction of plague and the disappearance of leprosy, knowing that these two diseases display great differences biologically, as well as clinically, epidemiologically, and historically. Leprosy is an exclusively human disease. It is chronic and progresses over many years; it is definitely horrible and disfiguring but rarely causes death. However, plague is a zoonosis that humans share with several animal species. It manifests as an acute and brutal fever; the sufferer's fate is determined in a few days, maybe only a few hours. Evidence suggests that leprosy is an ancient human disease, whereas plague is relatively recent. Indeed, the leprosy bacillus and humans seem better adapted to one another than do humans and the plague bacillus.

During certain plague epidemics, the passage from the bubonic form, transmitted by rats' fleas, to the pulmonary form, whose transmission is directly human to human, can be observed. Why did the disease not definitely evolve into a purely human form of attenuated virulence as, for example, measles? The violence of the pestilential waves across Europe during the Middle Ages can be surprising.[33] These plagues were unknown in classical antiquity and one would look in vain for descriptions in Hippocratic writings. A few plague epidemics may have erupted at the height of Rome, but they did not last long and left no site of infection on European territory. To trigger and maintain a plague epidemic, Yersin bacillus is necessary but not sufficient; a certain population density as well as the intervention of animal vectors are also indispensable. This is why pre-Roman Europe did not suffer from this ill. The first indisputable European pandemic of plague dates only from the sixth century (Justinian's Plague). After terrible ravages, it suddenly disappeared without diminishing its clinical manifestations. What was missing in Europe at that time was the animal population needed as a reservoir and vector. Plague is a disease of certain rodents where it has achieved biological equilibrium between host-parasite. Humans are only secondary hosts, subsidiary, and of little importance from a nonanthropocentric view. The epidemiological situation changed with the migration of the black rat, *Rattus rattus*, which invaded Europe during the later Middle Ages (according to Holsendorf's research, only in the twelfth century).[34] We can therefore understand why, after the second plague pandemic (with a culminating point in 1348), the disease could remain endemic: The rodent substrata and the denser human population assured its sustenance.

Having made this point, allow me to get back to the opening question: Why did the plague disappear after having reigned for several centuries as the uncompromising master? The usual explanations present no solid foundation.[35] Neither the changes in standard of living or hygiene, nor the immigration of the brown rat (which would have provoked the withdrawal of the black rat), nor quarantine measures, nor the questionable increase in human genetic resistance can justify the epidemiological situation in the nineteenth century. It is also said that the plague withdrew in the face of cholera. Besides the fact that the withdrawal of the plague quite markedly preceded the first apparition of cholera in Europe, no reasonable explanation has been given for the mechanism of this so-called antagonism. I nevertheless hope that the conjugated efforts of bacteriologists and historians will soon arrive at a new and better documented solution than those of the past. Therefore, research on phylogenic and immunologic relationships between plague bacillus and pseudo-tuberculosis bacillus (which could very well be a mutant of the first) promises interpretations rich in consequences for the historian.[36]

The antagonism between two diseases does not necessarily result from the antagonism between micro-organisms. It can sometimes derive from a conflict between the particular genetic state of the human organism and the germ of a given disease. In such cases, if the disease presents itself in endemic form, it exerts strong selective pressure in a way that, for physiologists, would seem abnormal. A good example is offered by the antagonism between malaria and sickle-cell anemia. The carriers of S and C hemoglobins (abnormal hemoglobins resulting from hereditary chromosomal abnormalities)[37] resist inoculation by hematozoa that provoke malaria, even though they are disadvantaged compared to normal individuals in favorable sanitary conditions. Therefore, this type of antagonism can favor hereditary abnormalities.[38] In fact, anemias in this group are practically unknown in Australia, which was never seriously exposed to malaria; by contrast, however, the areas of the world with the widest distribution of abnormal hemoglobin coincide with the ancient territories of endemic malaria.[39] Let us note in passing that the geography of blood diseases becomes more and more useful for the historian because it helps to identify ethnic units and uncover traces of human migrations.[40]

Antagonism between diseases can directly result from a very complicated sequence of the most diverse causes. Therefore, from the statistical viewpoint, notable antagonism exists between dysentery and coronary occlusion, or between typhoid fever and broncho-pulmonary cancer. Of course, there is no direct relationship between these diseases; their antagonism results only from the fact that the first characterizes the pathocenosis of a society with primitive hygiene, while the second dominates the pathocenosis of a technologically developed society. Diseases derived from "dirty hands" and polluted water are antagonists of diseases derived from "stress" or "degeneration" simply because they kill individuals before they can run the risks provoked by senescence.

In truth, there may be no direct antagonism or symbiosis in nature between two solitary and isolated diseases; but the real situation is always more complex. The historian of medicine confronts a dilemma similar to that of an astronomer observing the movements of at least three bodies. It is known that such problems cannot have theoretical solutions. Solutions must be found in an empirical manner. In our domain, this manner is represented by the quantitative study of pathocenosis.

It will have escaped no one's attention that the term *pathocenosis* has for a model the term *biocenosis*.[41] In fact, I drew as much inspiration as possible from the notions and methods elaborated by modern ecology. From a mathematical viewpoint, the study of the distribution of diseases by frequency poses a problem that corresponds to that of the distribution of animal and vegetal species determined by the number of living individuals in a biocenosis. Modern investigations[42] show that the distribution of species by quantitative importance corresponds to the normal log series (that is to say x; $\frac{x^2}{2}$; $\frac{x^3}{3}$; $\frac{x^4}{4}$; . . .).[43] Concerning the numerical analysis of diseases in human groups of relatively stable states of pathocenosis, I am still at the beginning of my research. Nevertheless, it would seem that here, too, certain regularities in distribution[44] and, perhaps, genuine laws of equilibrium will be found. The distribution of diseases as per their frequency seems to correspond to an interference between the simple logarithmic series and the normal log series.[45] Let us take note that the normal log series is to all phenomena of geometric progression what the Gauss curve is for all phenomena of arithmetic progression.

The birth of an epidemic marks a particular point that is situated outside the "normal" curve, but if the disease becomes endemic, it integrates within the normal distribution. However, it seems that this recuperation of a particular point is accomplished not only by the displacement of this point but also through change in the entire curve. This means that an epidemic ceases not only by the diminution of the number of persons suffering from that particular disease but by the simultaneous changes in frequency of a large number of other diseases. For the historian, such an observation has significant consequences.

To be sure, I hold no illusions on the difficulties that face quantitative analysis of pathocenosis. Three major obstacles are immediately encountered: the ambiguity of all definitions of morbid species, the practical difficulties of a correct diagnosis, and the impossibility of a complete inventory of diseases. These obstacles are important in our own time; they become insurmountable when looking back to the past. Only some principal trends, only some essential characteristics, of former pathocenoses can be studied. The historical documentation concerning this subject is more qualitative than quantitative, but I believe it to be sufficient at times to get a glimpse of, and to capture, certain structures or, as the English-speaking say, *patterns*.

Let me take as a first example the determination of the number of lepers. In the Middle Ages, this disease was poorly defined: The concept included true leprosy as well as psoriasis, tuberculosis of the skin, diverse other dermatosis, and even possibly, as believed by Holcomb[46] and Hudson,[47] syphilis. The diagnosis of leprosy was made in a manner similar to the judgment of a revolutionary tribunal, that is to say summarily and with a definite wish to declare all suspects guilty. The sources of error should not be overestimated. The paleopathological examination of bones found in ancient leper cemeteries confirms the diagnosis in a fairly large percentage of bodies.[48] The exact number of lepers in medieval Europe may not be known, but historical documentation is sufficient to establish relative quantity (at the height of this endemic disease, morbidity reached 1 to 5 percent). Let me now take an opposite example: Practically nothing is known about the number of people with cardiac diseases before the nineteenth century. The rare mentions of heart disease in ancient medical literature reflect the attitude of past official science and physicians' theoretical prejudices, not the real epidemiological situation.

A major difficulty in the quantitative study of past pathocenoses comes from the facts that the diseases themselves and their frequency change and physicians' ideas about disease change as well. The conceptual bases of medical diagnosis are far from immutable. The archives of the Académie nationale de médecine de Paris, for instance, have kept the material from an exceptionally rich and instructive study on morbidity in France from 1774 until 1794. On Vicq d'Azyr's (1748–1794) instigation, large quantities of documents were collected, but the analytical means of that time were insufficient to draw general conclusions. A team from the sixth section of the École pratique des hautes études is using these archives to construct a computer-assisted statistical analysis.[49] This endeavor is quite promising. Modern exploitation of this material, however, will require a preliminary analysis of the medical vocabulary. The names of the diseases have changed, and what is definitely worse, the system of references is no longer the same.

Physicians have turned from a clinical nosology based on external symptomatology and the notion of humors to an anatomical and etiological nosology. Therefore, the relationship between old and new terms is equivocal. How to translate "bilious fever" or "nervous fever" in contemporary terminology? The eighteenth century's "morbus cholera" has nothing to do with today's cholera but corresponds to dysentery and a few similar diseases. Historically, "scurvy" is a complex of diseases where avitaminosis is but one of the pathogenic factors.[50] Very often, an ancient diagnosis would encompass several diseases in today's sense. Hence, the eighteenth century's "phthisis" is not necessarily today's pulmonary tuberculosis. The risk of a vicious circle seems considerable to me. In fact, "translation" is often possible only through consideration of the frequency of certain symptoms in today's clinical framework. Apparently justified, this effort nevertheless will infuse our conclusions with a statistical frequency, which should precisely result from historical facts and not be attributed to them.

The study of past pathocenoses also faces an obstacle that historians of everyday life know well: Historical sources frequently leave out the most common occurrences. The historian or the chronicler notices epidemics because they are extraordinary, whereas certain endemic diseases, which act surreptitiously, become "historically invisible" because of their extent and commonness. By studying the dynamics of pathocenosis, the medical historian's attention will certainly be drawn to two very prominent phases: on the one hand, toward periods of equilibrium of pathocenosis, and on the other, toward periods of upheaval. In a future article, I will try to show that the principal tasks of future historians of diseases will be to establish the various types of pathocenotic equilibrium and to study the factors that can hinder, and finally provoke, the rupture of this equilibrium.[51]

The Concept of Emerging Disease

During my research on the history of AIDS, I had to examine the problem of diseases considered "new." From the first clinical and epidemiological definition of AIDS, the general public and even physicians have questioned the ancient quality of this plague. It quickly becomes clear that in a certain sense AIDS is a "new" disease, while in another it is not.[1] AIDS appeared as a new disease to the extent that such a pathological state could not even be conceptualized as a nosological entity prior to the 1970s. Back then, a disease was defined by its symptoms or by lesions in anatomical structures. However, neither were applicable to AIDS, which was a disease without typical clinical symptoms, a disease characterized by subcellular and invisible symptoms, and provoked by a germ that was undetectable prior to the recent use of certain analytical tools.

AIDS is certainly a new disease with respect to its pandemic dimension, but this does not necessarily entail that the HIV-1 and HIV-2 viruses are, in a strict sense, newcomers, mutants whose ancestors were never pathogenic. It seems increasingly clear that AIDS is not a new disease in the strong sense of the term; HIV retroviruses have existed for quite a long time, provoking—

behind the mask of other infectious diseases—sporadic pathological or even collective states, but limited in time and space.

To answer the question concerning the "novelty" of AIDS correctly, the usual terminology must be revised and defined with greater precision. What does the expression "new disease" mean exactly? Is a disease new because physicians have not recognized it prior to a given date, or is it new in the sense that it did not exist "in reality"? And in the latter case, is the disease new only in a certain place or new for the whole world? Is it new only with respect to the recent past or new when considering the entire history of mankind?

A change in vocabulary is crucial to avoid confused ideas and misunderstandings. The notion of "novelty" must therefore be replaced with that of "emergence." The epithet "emergent" can unambiguously apply to all diseases perceived as "new" at a given time and in a given population, while the expression "new" is restricted to diseases that, prior to a specific moment, have not existed in a population as clinical realities. At the outset, I should state that my remarks in this paper will be limited to infectious diseases or, more precisely, to the pathological states caused by microbes. These are diseases that change so much with time that one can say, to echo Charles Nicolle (1866–1936), that they are born and that they die.[2]

The Ancient Quality of the Notion of "New Disease" and the Novelty of the Notion of "Emerging Disease"

The notion of "new disease" is itself very old and can be found in a number of ancient texts. For instance, the oracular decrees of Pharaonic Egypt mention "all diseases from books" and "every disease about which we are not informed"—that is, diseases for which "technical knowledge" was unavailable. Doctors in ancient Egypt crafted the concept of diseases not yet encountered or, rather, not yet "invented" by hostile divine forces. As a new stratagem in the hands of these forces, these diseases were created to evade ordinary treatments and to be more harmful than older ones. In addition to this strong sense, physicians in ancient Egypt also had in mind a weaker sense of the notion of new disease: A disease could be new only because it was imported from abroad.[3]

In Europe, the pestilence that descended on the Attic population in 430 BC was the first serious epidemic of a collective disease with a high mortality rate for which the symptoms were known. It must be emphasized that this epidemic was understood at the time as a novel phenomenon. In his masterly description of this plague, historian Thucydides (465–395 BC) noted that "no pestilence of such extent nor any scourge so destructive to human lives is on record anywhere."[4] Doctors were unable to cope with the disease, he continued, "since they at first had to treat it without knowing its nature."[5]

Experts never quite reached a consensus with respect to the enigmatic character of this epidemic. In my opinion, the pestilence of Athens was exanthematic typhus striking an immunologically virgin soil, combined with nutritional disorders and secondary infections. Be that as it may, the disease was perhaps new to Greece, as Athenians thought it was, but it was certainly not new for the whole of humanity. Thucydides reported that "the disease began, it is said, in Ethiopia beyond Egypt."[6] Since antiquity, Africa was reputed to be the cradle of pestilence and of the majority of diseases taken to be new.

Following the unification of Mediterranean countries under Roman reign, several diseases not found in earlier medical treatises began to spread during the first centuries of our era. Writing about such "new diseases," Pliny the Elder (23–79) and Seneca (c. 1–65) attributed their appearance in Europe to the mixing of populations and to changes in lifestyle, in particular stressing the intemperance of the new ruling class in Rome.[7] Plutarch (46–125) spoke of this problem in terms of a philosophical aporia: Either these new diseases have truly appeared for the first time, and one is thus forced to admit that the inner nature of things can change, or all diseases have always existed, but their emergence has escaped the attention of physicians. Leaning toward the second position, Plutarch believed that some diseases seem new because they come from distant countries, perhaps even from an outside world.[8] Democritus (c. 460–370 BC) would have already professed that some diseases are of extraterrestrial origin and have resulted from the transfer of dangerous particles by cosmic rays.[9]

For physicians in the Middle Ages, the irruption of the "Black Death" in the fourteenth century was extraordinarily surprising and was comparable to the emergence of syphilis at the very end of the fifteenth century. Given that Hippocrates's (c. 460–370 BC) and Galen's (129–217/216)[10] writings contain no information about such ailments, their novelty had to be acknowledged and proper descriptions furnished. But it is above all during the eighteenth century that European doctors observed and described a large number of diseases that seemed new to them. To explain the unexpected rise of those new nosological entities, seventeenth- and eighteenth-century scholars called attention to several factors, including: the intrinsic variability of diseases, the new astral conjunctions, the population increase, the intensification of contacts with non-European races, and the degeneration of mankind in general and of the white race in particular.

The appearance of several "new" diseases, most of which were not particularly serious, was due, on the one hand, to significant changes in living conditions and consecutive upheavals in the structure of morbidity, and, on the other hand, to a new clinical conceptualization of the medical model of disease.[11] The same process occurred in the nineteenth century, first with the development of the anatomo-clinical method and then with the rise of microbiology, facilitated by the development of urbanization and the industrial revolution. With the advances of biological knowledge in the twentieth century—notably the rise of medical imaging and biochemical and immunological tests, accompanied by new conceptions that situate the basic lesion at a cellular level—we witness the discovery, or rather the creation, of nosological entities that could not have even been imagined in the past.[12]

An Attempt at Classification

While analyzing recent medico-historical literature as well as ancient publications that signal the appearance of a scourge considered "new" or describing a previously unknown clinical entity, I noted that a disease can be classified as emergent in at least five different historical situations:

1. It existed before it was first described, but it escaped the medical gaze because it could not be conceptualized as a nosological entity;
2. It existed but was not noticed until a quantitative and/or qualitative change in its manifestations occurred;
3. It did not exist in a particular region of the world before its introduction from other regions;
4. It never existed in a human population but only in an animal population;
5. It is completely new—the triggering germ and/or the necessary environmental conditions did not exist prior to the first clinical manifestations.

The theoretical possibility that a disease is artificially produced should also be considered. That is to say, its emergence could be due to an intentional act (as in preparation for biological warfare) or to an unforeseen accident (for instance, in the context of biotechnological manipulations). Such diseases would then represent a subclass of the category of absolutely new diseases. If by definition one can, and even should, designate as "new" all diseases in the fifth category, this adjective should only apply to diseases in the fourth category with an important caveat: "New" in the strong sense of the term can only be a human disease of an interspecific origin that results from the very first pathogenic passage from germs to humans. Even if the ensuing epidemic that results from this initial passage is completely extinguished, the reappearance of the same human disease from an animal source is not a new disease in the strong sense but only a case of emergence.

This classification of disease emergence in one of the five aforementioned categories is not absolute but relative to a given historical situation. The same disease can be endemic in a country and emergent in another; alternatively, in a given population, a disease can be emergent following its passage from animal to man at a specific historical moment or after a change in the ecosystem at another moment.

Category 1: The Emergence of Nosological Entities as Concepts

Contemporary English-speaking scholars make a distinction between two aspects of the concept of disease and attribute different meaning to the words *illness* (which designates the illness as it is experienced by a sick person or as it is perceived by their relatives) and *disease* (which is a concept constructed in a nosological system). A "disease" (broadly construed) can emerge as an unexpected novelty in terms of suffering, both at the individual and the collective levels, and in terms of its "objective cause" and scientific explanation ("the medical model of disease").

Among the diseases considered new by physicians in the Renaissance, some (such as exanthematic typhus, mumps, and measles) were clearly not. The clinical picture of the first two was clearly described in medical texts dating from antiquity.[13] This was not the case for measles, although this disease was caused by the same virus as zona, which was quite a well-known disease to ancient Greek physicians. Like many other childhood diseases, measles probably went unnoticed or was mistaken for some other acute exanthemas in children.[14] But the possibility that the virus in question was primarily, or perhaps even exclusively, the cause of zona cannot be ruled out.

A considerable number of then-unknown diseases were discovered or intellectually produced during the "biomedical revolutions" of the mid-seventeenth, nineteenth, and twentieth centuries. Neo-Hippocratic physicians of the seventeenth century, above all Thomas Sydenham (1624–1689), created nosological entities on the basis of precise clinical distinctions. Those diseases were, of course, falsely new; long prevalent, they had only recently become "visible" and accessible to the medical gaze. Throughout the first half of the nineteenth century, other falsely new diseases appeared following a similar process thanks to the systematic application of the anatomo-clinical method and, closer to the present time, following the triumph of cellular pathology, medical bacteriology, and the beginning of molecular biology.

Category 2: The Emergence of Diseases after a Quantitative or Qualitative Change in Their Manifestations

Contrary to a widely held but naïve opinion, a pathogenic germ is not the cause of an epidemic. A specific microbe is not the cause of a disease (in the strong sense) at the collective level; it is a necessary condition, but it is not sufficient. The introduction and the persistence of a disease within a population also depend on other factors, which are both biological (the immune state of the host, vectors, and concomitant germs) and social (particularities of behaviors, standards of living, etc.). Microbiological phenomena, such as mutation of pathogenic germs and immunological phenomena in human bodies, do play an undeniable role in the historical transformation of diseases, but this should not lead us to ignore the decisive impact of human behavior on such processes. Agriculture, animal husbandry, changes in the milieu, mixing of populations, and changes in habits have undoubtedly exerted a stronger influence on outbreaks and epidemic diseases than modifications in microbial genomes. Epidemics, indeed, are almost always the result of new opportunities offered to ancient germs by human actions, not the result of new germs as such.[15] Because of the cyclical aspect of their manifestations, epidemic diseases, such as influenza, dengue, and even smallpox, as well as endemic diseases, such as malaria (notably the falciparum type), have sometimes been experienced as new plagues by both laypeople and physicians.[16]

At least since the end of the Middle Ages, influenza has returned fairly regularly and at determinate intervals because of a very particular process of genetic recombination that evades the human body's immunological defenses in a cyclical manner. The two antigens attached to the surface of the influenza virus (hemagglutinin and neuraminidase) change continuously; following small changes in the viral genome that determine the expression of these surface antigens and that give rise to epidemic waves of low amplitude (antigenic *drift*), other, more radical changes (antigenic *shift*) occur and provide germs with a "new" outward appearance that elicits the sudden emergence of acute epidemics two or three times each century.[17] Hemorrhagic dengue fever, for example, periodically returns thanks to a recombination process that seems to operate during an infection with two or more types of arboviruses.[18]

The emergence of other diseases may be due to the synergic action of germs that are quite different. Our knowledge on this subject is still limited because microbiological research was long dominated by the methodological principle of specificity ("one disease–one germ").

Arthropod-transmissible diseases (malaria, yellow fever, sleeping sickness, rickettsia, and others) can emerge because of changes in the geographical distribution of vectors. Acute effects can occur not only after significant changes in the ecosystem of these vectors, such as deforestation or water development projects, but also, and in an unexpected manner, by apparently minor variations in human behavior.

Writers of ancient medical texts had intimate knowledge of the clinical picture of diphtheria (only described as a nosological entity during its heightened epidemic in the nineteenth century) and erythematous angina pultaceous; they were unaware of the clinical picture of scarlet fever. This disease did not manifest itself in its typical form until the end of the Middle Ages, although in all likelihood the germ already existed. It is only in the recent past that a large number of serious infectious diseases, unknown until then, were described, even though, in principle, we had the means to diagnose them. Sometimes, serious illnesses could result from a germ's mutation, and could thus be cases of new diseases in the strong sense, although most of the time the causal agent already existed prior to the epidemic; the visible flare-up of this ailment was then due to purely socio-ecological factors.[19] Toxic shock syndrome (TSS) and "Legionnaires' disease" are particularly instructive examples in this respect. Described by American researchers in 1978 as a "new" feminine disease, toxic shock syndrome is a singular expression of the old and ubiquitous golden staphylococcus bacterium.[20] TSS's emergence results both from particular aspects of modern women's behavior and from a change in the biology of the germ that made it more virulent.

"Legionnaires' disease" is a form of pneumonia with an unusual profile that infected several participants at the meeting of the American Legion in Pennsylvania, held at a luxurious hotel in Philadelphia in 1976. An exemplary epidemiological investigation led by the Centers for Disease Control in Atlanta rapidly established that the responsible microbe was a bacterium unknown until then. Named *Legionella pneumophila*, this bacterium, it turns out, was able to cause an affliction that looked exactly like a new nosological entity. Further microbiological investigations, however, demonstrated that some species of *Legionella* had long existed in our environment and were even widespread. These are relatively innocuous germs. Philadelphia's epidemic, therefore, was not due to a mutation in a germ that led to a particularly virulent bacterial strain but rather to a new intervening ecological factor. Strains of *Legionella* turned deadly because they were allowed to multiply in stagnant water inside an air-conditioning system and were distributed by aerosol.[21]

Other small-scale epidemics of this disease followed in the wake of the outbreak in Philadelphia, always in places where sanitary systems were quite sophisticated. These cases illustrate the ways in which technical development can facilitate diseases, not only by degrading the natural environment with polluting agents but also by generating harmful effects resulting from purportedly more "hygienic" living conditions. What should be kept in mind from this example is that earlier cases of this disease classified as atypical viral pneumonia could now retroactively be diagnosed as Legionnaires' disease. In my opinion, most iatrogenic emerging diseases, such as inoculate hepatitis, should be placed in this category.

Category 3: The Emergence of Diseases through Their Introduction from Another Region

The most important of the "new" diseases discussed by Pliny the Elder and Celsus (c. 25–50) is leprosy (which they called *elephantiasis*). It is truly an ancient disease: Evolutionarily, its germ probably originates from the human branch of mycobacterium. From time immemorial, leprosy spread across India and China. It also struck Mesopotamia, but the chain of infection broke down in Europe and the disease probably spared pre-Columbian America. In Greece, leprosy was known only in the Hippocratic era because of rare, imported cases. First only sporadic in the Mediterranean east, leprosy expanded during the Hellenistic period and was introduced into Western Europe at the height of the Roman Empire.[22]

The most illuminating examples of this category pertain to the exchange of microbes between the Two Worlds: the arrival of venereal syphilis in Europe and the arrival of smallpox and a deadly form of influenza in America. The history of syphilis is fairly complex and cannot be reduced to a simple transmission of germs that did not previously exist in Europe. Not only did pathogenic Treponema exist in the ancient world as early as the Paleolithic era, but human-adapted strains also probably circulated long before the 1494 epidemic. However, it is no less true that the "great pox" essentially resulted from the introduction of a particular mutant form of the American germ into European countries.[23] The extraordinary development of techniques of biological analysis, such as genetic amplification (PCR), opens up new perspectives for historians of diseases. We have identified specific antibodies for syphilis in the mummy of Maria d'Aragona (1503–1568), who died in Naples in 1568; and we can hope to find and to analyze the Treponema genome of the sixteenth century.[24]

American syphilis itself might be of African descent. Native and inapparent (or at least less virulent) forms of Treponema exist among baboons and gorillas in central Africa. André Fribourg-Blanc (1888–1963), Henri H. Mollaret (1923–2008), and others have shown the presence of anti-Treponema antibodies in the blood and Treponema in the popliteal lymph nodes in otherwise apparently healthy West African cynocephali monkeys.[25] The affliction, thus, probably jumped from monkey to man in Africa or, more likely, the common ancestor of humans and high apes was already parasitized by paleo-Treponema. Millennia later, the pathogenic germ returned to the African continent from America in a much more virulent form.[26]

Without knowing it, European conquerors carried out a true biological war. The viruses they transported within their own bodies or those of their slaves and domestic animals were indeed powerful allies.[27] The effects of smallpox and measles—new diseases for the Native American population— were dramatic.[28] Other viruses acted more subtly but their selective impact on indigenous populations had equally severe consequences, both from a military and a demographic point of view. A virus hitherto absent on the American continent—perhaps the germ for swine flu—produced devastating epidemics.[29]

Smallpox was introduced in America by black slaves at the beginning of the sixteenth century. Evidence strongly suggests, however, that the first infection came from Western Africa and not from Europe. Some historians have argued that was a benign form (*variola minor*) during Middle Ages in Europe and Renaissance smallpox and that the virulent form of the disease (*variola major*) did not exist on the European continent until the end of the sixteenth century. Smallpox, they claim, might have been introduced via a new strain either directly from Africa or from an American intermediary.[30] But even if this view were correct, it would only be a case of reintroduction because a serious smallpox epidemic struck the Roman Empire at least as far back as Galen's time.

The origin of yellow fever remains an open question. Although some consider that it was endemic in American native populations, strong arguments favor the hypothesis that it was introduced in America after the conquest. By accepting the second hypothesis, however, one must also admit that the disease came not from Europe but from Africa.[31] At any rate, this leads to the epidemiological events of the nineteenth century that were triggered by the colonial expansion of powerful European countries, which must be held responsible for the introduction of yellow fever and many other so-called tropical diseases in countries that did not previously suffer from them.[32]

Some diseases that have an exclusively human causative agent cannot persist in populations where the number of susceptible individuals fails to reach a critical threshold. Such cases essentially concern acute viral diseases that swiftly lead either to the patient's death or to his full recovery. The disappearance of these diseases from a population is often followed by the dramatic reintroduction of the same pathogenic agent. The history of measles provides an instructive case of a disease that retreated at the local level and returned later, with dramatic consequences, at the collective level.[33]

Category 4: Disease Emergence via the Passage of Germs from Animals to Man

Many diseases caused by "horizontally" transmitted germs from domestic animals to humans during the Neolithic period can be understood as genuinely "new." Tuberculosis and acute eruptive fever caused by poxviruses are among the most important in this category.

Tuberculosis is caused by a mycobacterium of extra-human origin that likely emerged in man in two steps: The bovine strain first adapted to humans and then mutated into a specifically human strain. A very large time gap possibly separates the two events.[34]

Man is unlikely to be the primary host of brucellosis (Malta fever), an endemic-epidemic bacterial disease, though very little is known about the history and the conditions leading up to the transfer of this bacterium from livestock to human populations. The distribution of brucellosis in domestic animals still supports the hypothesis that it first appeared in the Neolithic period.[35]

Like measles, whose virus is related to bovine "plague" and canine distemper, smallpox probably dates to the Neolithic period. These viruses were not born in Europe but were imported into the Mediterranean region from Africa or Asia.[36] Smallpox is—or, rather, was—a strictly human disease, even if viruses belonging to the same group can also infect numerous domestic animals. The closest relative of the human poxvirus is a pathogenic germ found among small African monkeys.

Influenza viruses possibly originate from a strain that was once a parasite in one or several animal species, notably aquatic birds. We do not know when the virus adapted to humans, but it is clear that influenza remains a zoonosis and that the genetic recombination, responsible for the cyclical appearance of epidemics in human populations, occurs in animal bodies and most likely in pigs. Note that influenza does not affect monkeys, which seems to indicate that it also spared prehistoric human populations.[37]

The plague itself did not strike Europe before the end of antiquity. It is even quite likely that it did not exist in any part of the world as a human epidemic disease before the historical age. Plague is a disease in the rodent population that affects humans only accidentally. The first probable reference to a plague epidemic goes back to the third century BC in Africa. The disease struck three times, each time resulting in devastating pandemics with no immediate precursors: the so-called Justinian Plague in the sixth century (that allegedly began in Egypt); the "Black Death" in the fourteenth century (the most deadly pandemic of all time, originating from China); and the Manchurian Plague in the nineteenth century.[38] We now distinguish three varieties of *Yersinia pestis* found in nature that are, according to the attractive hypothesis proposed by R. Devignat, descendants of the germs responsible for the three great pandemics.[39]

The biological and ecological conditions that allowed plague viruses to go from rodents to humans remain in large part unknown. The mutation of *Yersinia pestis* at a single genetic site, or at two sites of its close cousin, *Yersinia pseudo-tuberculosis*, radically alters the virulence of a strain. It is conceivable that a *longue durée* enzootic plague among rats and sylvatic rodents resulted from the predominance of low-virulence strains favored by natural selection.

In other words, mutations at critical loci could have resulted in the emergence of hypervirulent strains. When passed on to humans (under ecological circumstances still poorly understood), these could provoke a disastrous epidemic.[40]

Viruses associated with African fevers, named by their place names (Marburg, Lassa, and Ebola), are new pathogenic factors whose emergence occurred between 1967 and 1976 and that were unknown until then. These diseases are not really new but present themselves as such only because they broke out from their isolated ecological niches. Monkeys (for Marburg and Ebola fevers) and sylvatic rodents (for Lassa fever) are the natural hosts of these viruses, but their passage to humans does not itself seem to be a new event. Retrospective serological investigations have shown that these viruses already existed prior to epidemics that made them known. These cases demonstrate the existence of silent, endemic viruses with potential for interhuman transmission in Africa, though specific circumstances are required for them to emerge as disease agents.[41]

In a sense, twentieth-century African fevers remind us of the famous "English sweating sickness" that appeared around 1480 in England. After erupting in limited outbreaks within Continental Europe and causing disease until the mid-sixteenth century, the disease vanished rapidly and completely. Its emergence and its disappearance still remain equally enigmatic. Evidence suggests that the English sweating sickness was caused by a virus, even if no one could either identify it or confidently connect it with pathological manifestations observed in other eras or in other parts of the world.[42]

Category 5: The Emergence of New Diseases in the Strong Sense

A certain continuity with the past exists even in the emergence of this last category because the pathogenic germ cannot result from a spontaneous generation. A new disease necessarily results from the adaptive transformation of a saprophyte germ or, as is often the case, from the transformation of a commensal germ in humans or animals. Such transformations are both uncommon and difficult to establish.

Among diseases considered new at the beginning of our century, the most enigmatic is lethargic encephalitis. Described in 1917 by Jean-René Cruchet (1875–1959) in Paris and by Constantin von Economo (1876–1931) in Vienna, this disease had appeared in the form of an epidemic around 1915 in

China and then it ravaged the entire world; it caused approximately half a million deaths over a decade. Lethargic encephalitis, however, seems to have totally disappeared after 1930. Although no one could identify the causative agent, evidence points to a mutated form of the influenza virus and suggests that its descendants might still live as parasites on animals.[43]

The most instructive historical case of a new disease in the strong sense is perhaps the emergence of cholera in the nineteenth century. Starting in 1817, seven pandemic waves of cholera swept almost every single country in the world. Prior to 1830, however, cholera did not even exist in Europe and its arrival from the Ganges delta is well established. From a European or an American point of view, cholera falls into the third category of emergence (introduction from a foreign country). But should we continue to accept the widely held opinion that cholera existed in India since the beginning of time? Recent research currently favors the hypothesis that cholera was a genuinely new disease. It emerged in India as a serious infectious disease only at the end of the eighteenth or the beginning of the nineteenth century via the genetic transformation of a saprophytic vibrio.[44]

To conclude, I can but briefly indicate two directions of research in epidemiology that remain insufficiently explored, although both deserve our full attention. On the one hand, these research avenues concern the global study of morbidity in a population, on the other hand, the projection of our historical knowledge into the future.

Examining the interaction between diseases is critical to understanding the suppression as well as the emergence of pathological conditions. To this end, the notion of *pathocenosis* can help us gain a better understanding of the relationships between certain pairs of diseases (e.g., between leprosy and tuberculosis and between tuberculosis and AIDS) and, above all, better grasp the natural laws governing the relationships between diseases in a given population at a given historical moment.[45]

Cicero's (106–43 BC) famous dictum—*Historia magistra vitae*[46]—has but a limited and pragmatic value for a historian of diseases. In the field of epidemiology, if history does not in fact teach us what we must do in the present in order to ensure a better future, it nonetheless allows us to better understand what appears before our very eyes and can put us on guard against nefarious actions. The desire to place the study of the past in the service of the future, in particular, inspires certain current research on "emerging viruses."[47] The smallest changes in our milieu and in our habits can lead pathogenic germs to pass from animal to man, and the less pathogenic or even completely inactive viruses, hidden within the human body, to reappear as serious troublemakers.[48] We are definitely not protected from mutations or nefarious adaptations of normally nonpathogenic viruses that abound in nature, particularly in aquatic environments.[49] Perhaps we shall be able to predict, however vaguely, some future epidemiological events.

THREE

Some Unorthodox Views and a Selection
Hypothesis on the Origin of the AIDS Viruses

In speaking of the beginning of AIDS, or of the first stages in the history of any emerging infectious disease,[1] it is necessary to discriminate and clearly delineate three problems: the origin of the germ; the first appearance of the disease in sporadic form; and the onset of the epidemic. The emergence of AIDS is a historical process in three superimposed stages, each provoked by particular causes.

In my book *History of AIDS*,[2] I traced the chronology of our growing awareness of these problems: first, during the summer of 1981, the declaration of a specific epidemic; then, one year later, the definition and naming of a previously unknown disease, the earliest hypotheses concerning its beginning; and, finally, after isolation and identification of the causative virus in 1983, the attempts to establish a phylogenetic tree of retroviruses and to explain the origin of virulent strains of the AIDS lentiviruses. I shall discuss here only the last point, the emergence of highly pathogenic strains, which in fact necessarily preceded other events related to this epidemic.

No pathogenic virus is entirely new, nor can it emerge ex nihilo. It comes from an ancestor that must have possessed neighboring genetic characteristics and also must have persisted somewhere in animal or human populations. It is possible, even probable, that this ancestor was harmless or only slightly pathogenic for its original host population. All the scenarios and hypotheses of the origin of the AIDS virus that I have encountered in the literature, and those that I have managed to think of in addition, are presented in an accompanying table (Table 1). I cannot discuss them all in this paper.

The viral cause of AIDS was suspected from the moment that its transmission by filtered blood products was recognized (specifically, after the publication in July 1982 of three cases in hemophiliacs). Following the isolation of the LAV virus by Luc Montagnier's team in Paris in January 1983 and one year later of the HTLV–III virus by Robert Gallo's team in Bethesda, the scientific community gradually began to realize that these were two samples of a single virus (now called HIV—an acronym adopted in 1986 to signify Human Immunodeficiency Virus)[3] and that this specific virus, from the lentiviral branch of retroviruses, is the "cause" of AIDS. Later, the group at the Institut Pasteur isolated from the lymphocytes of a West African patient another virus with morphological and genetic characteristics similar to the first. Less virulent than the original HIV, this HIV–2 is also recognized as a cause of the acquired immunodeficiency syndrome.

Is HIV the Cause of AIDS?

There are still a few competent scientists who do not accept the HIV retrovirus as the "cause" of AIDS. Thus Peter Duesberg (b. 1936), professor of molecular and cellular biology at Berkeley and recognized oncogene specialist, argues against this commonly accepted opinion and refuses to see in HIV anything more than a harmless or only slightly pathogenic opportunistic organism, which occurs frequently in the conditions that favor the true cause.[4] Duesberg declares himself ready to be injected with an HIV preparation on the proviso that it has been truly purified and free of all suspicion of contamination. Of course, as this suspicion cannot be eliminated, Duesberg never tried to perform something analogous to the famous experiment of Max von Pettenkofer (1818–1901) who, in his dispute with Robert Koch (1843–1910), drank a broth of culture containing cholera vibrios.

Duesberg's arguments are based on statistical and methodological considerations and not on experimental proofs or counterproofs. At the outset, they were not without value, because they delineated murky areas in accepted

TABLE 1. Possible explanations of the origin of the human AIDS virus

---

Part A. Improbable explanations

(I)   HIV lentiretroviruses are old and widespread but act only as opportunistic companions of AIDS, the cause of which is multifactorial

(II)   HIV–1 was artificially made by humans
        (1)   deliberately
        (2)   accidentally

(III)   HIV, or its immediate ancestor, came from outer space

(IV)   HIV came from viruses of plants or not-primate animals

Part B. Probable explanations

(V)   HIV came from primate viruses recently (cross-transmission by accidental contact with simian blood or by some particular medical activity)
        (1)   Simian lentiretroviruses were already pathogenic for humans
        (2)   Simian lentiretroviruses were precursors that adapted to humans to cause AIDS

(VI)   HIV existed in humans for a long time in virulent forms
        (1)   One or few isolated populations
        (2)   In diverse populations but inducing only sporadic cases of AIDS

(VII)   HIV existed for a long time in humans in harmless or only slightly pathogenic forms that became virulent
        (1)   Through a single biological event (mutation)
        (2)   Through an encounter with a particular cofactor
        (3)   By natural selection (under the pressure of social and biological events that facilitate infection)

---

opinions and demonstrated that certain important problems had not yet been satisfactorily resolved. But Duesberg leapt too rapidly from explanatory difficulties to completely negative conclusions. His scientific jousting might be amusing if the stakes were not so high and the practical consequences not so serious. It shall not be forgotten that one of the corollaries of the Duesberg hypothesis is that AIDS cannot be transmitted by either homosexual or bisexual activity.[5]

While the "strong" criticism of HIV now seems to have been definitively rebutted, a "weak" form of the criticism still remains viable. Can the AIDS retrovirus act alone or does it require obligatory accomplices? Numerous observations suggest the existence of cofactors, such as mycoplasma, or other germs and peculiar conditions.

The "central paradox" of the HIV hypothesis is that relatively few lymphocytes are infected considering the massive functional deficit and progressive cell loss. The most satisfying responses to this objection presume the existence of an autoimmune component in the pathogenesis of AIDS. In persons infected by HIV, as opposed to those who are not, numerous uninfected CD4 cells are programmed for cell death by apoptosis, a genetically determined autodestructive process defined by oncologists in the early 1970s.[6]

The pathophysiological mechanism of the generalized weight loss in AIDS patients ("wasting syndrome" or AIDS cachexia) is still enigmatic.[7] Most likely these perturbations in metabolism are also linked to autoimmune reactions. Jacques Leibowitch suggests that this marasmus might be a sort of collective suicide led by the ameboid mesodermal cells, which Metchnikoff (1845–1916) saw as being actively involved in morphogenesis, senility, and atrophy.[8]

AIDS could be interpreted as a syndrome essentially characterized by the autodestruction of the immune system and a nonspecific suicidal reaction of the organism. HIV might set off a "bomb," a programmed molecular mechanism for self-destruction, that we carry in the deepest recesses of our bodies. If this is the case, then the human immunodeficiency viruses are only the trigger and not the efficient cause of a deadly disease. Even so, HIV–1 and HIV–2 undoubtedly provoke AIDS and must be held responsible for the present epidemics.

There are some patients with symptoms that correspond to the clinical definition of AIDS in whom no trace of the known AIDS viruses can be found. Until 1992, specialists found these cases to be of only limited interest, since imperfections in the methods of viral detection meant that definitive exclusion of HIV infection was not possible. The existence of "AIDS without HIV" is now generally acknowledged, but all known cases are sporadic and completely unrelated to each other. They correspond to a clinical definition of AIDS, but, taken together, they constitute a heterogeneous group, a syndrome. American scientists dubbed cases of "AIDS without virus" as ICL (idiopathic CD4 lymphocytopenia). This name sounds familiar to a medical historian: In the past, doctors commonly used the adjective "idiopathic" to dissimulate their ignorance. Nothing proves that ICL is new. It has surfaced now just because of the more sophisticated means of detection and of a tremendous increase in the surveillance of immune disorders.[9]

The Subversive Origin of HIV

The hypothesis that the AIDS virus might have been produced deliberately or accidentally by genetic engineering has no scientific basis, but the circumstances of its formulation and dissemination in the press deserves consideration by historians.[10]

On October 30, 1985, the Soviet magazine *Literaturnaya Gazeta* published a sensational article claiming that the AIDS virus was a biological weapon produced in an American laboratory.[11] The author, Russian journalist V. Zapevalov, cited revelations made earlier the same month by *The Patriot*, a Delhi newspaper influenced by Communists. According to Zapevalov, the virus had been brought from the United States to Zaire (the former Belgian Congo) in 1978 by American doctors and used to vaccinate indigent patients, either intentionally or through a technical error. Thus, the epidemic began in central Africa. The almost simultaneous appearance of an epidemic focus in America could be explained by the negligence of the researchers who allowed the lethal virus to escape from the military laboratory where it had been created by genetic manipulation.

Following a well-established method of media intoxication, the Soviet journalists claimed to know nothing from firsthand sources and cited only items in the foreign press. Who outside of India was aware of the political connections of *The Patriot*? Furthermore, the most amusing aspect of the affair is that an on-the-spot inquiry made a year later showed that the news item in question had never been published in the quoted Indian newspaper.[12] Nevertheless, the rumor circulated, sustained by press releases. On October 31, 1986, *Pravda*, the official newspaper of the Communist Party of the Soviet Union, published a cartoon showing a doctor in a white lab coat offering a test to an American officer with one hand and accepting a wad of money with the other. The caption said: "The AIDS virus, a terrible disease for which up to now no known cure has been found, was, in the opinion of some Western researchers, created in the laboratories of the Pentagon."[13]

By September 1986, during the summit of nonaligned nations at the Harare, Zimbabwe, the accusation had taken on serious political overtones. All delegates received an official-looking report signed by "doctor J. and L. Segal, researchers at the Institut Pasteur," and by a scientist, R. Dehmlow, whose name suggested a Slavic origin. The report claimed that the AIDS virus could have no origin other than genetic manipulation. Taking into consideration the incubation period, the first appearance of the disease coincided exactly with the opening of the P4 laboratory at Fort Detrick, Maryland. The global dissemination of AIDS began from New York City, not far from Fort Detrick. Therefore, the report concluded, the AIDS virus is artificial and its production is linked to the preparations for biological warfare.[14]

Investigations made after the Harare summit revealed that the first two signatories, Jacob and Lilli Segal, were not researchers at the Institut Pasteur but teachers in East Berlin. They handed the African politicians a text garnished with untestable suppositions, outright lies, and scientific impossibilities, though carefully composed for precise political ends. Devoid of any proof, the accusations turned on a few pieces of circumstantial evidence: First, in 1980 there had been rumors that the US Navy was experimenting with biological weaponry designed to work specifically against black people (a so-called ethnic weapon); second, in the critical years, the US Defense Department had been increasing its funding for biological research; and third, the main chemical and biological warfare facilities were located at Fort Detrick, only a few miles from the NIH laboratories at Bethesda, where Robert Gallo was working on retroviruses.

Indeed, Fort Detrick is not far from Gallo's place of work, and, without any written details, it was claimed that he was posing as the "discoverer" of a virus of which he was, more precisely, one of the creators. In his memoirs, Gallo said the following:

> In East Berlin [in 1988] I attempted, unsuccessfully, to meet a Professor Siegel, who had invented and actively promoted the notion that HIV was created by a U.S. government scientist, with the help of the Frederick Cancer Research Facility in Maryland. . . . I am told that Siegel's thesis was that HIV is a recombination of HTLV–1 with a lentiretrovirus of sheep known as the visna virus, constructed by recombinant DNA technology. This is an interesting notion that superficially makes tantalizing science sense, since the HTLV component would conceivably provide the human and T-cell targeting capacity, while the visna component would give it its macrophage tropism and its ability to mutate so easily. The problem is that HIV has been proven to have existed long before recombinant DNA technology was invented; moreover, there was little relationship between the published gene sequence of HTLV–1 and HIV, and only a small amount of genetic similarity between the gene sequence of visna and HIV. The notion was simply stupid, malicious, or both.[15]

Spreading from the Soviet Union and East Germany, the speculative calumny was very popular in Africa.[16] It was taken and even elaborated in Occidental countries, above all by an English doctor, John Seale, and by a French journalist, Rolande Girard. Seale, a noted specialist in venereal disease in London, alerted both the medical community (through an editorial published in the official journal of the Royal Society of Medicine)[17] and the public (through an interview that appeared in a widely circulated London weekly).[18] The idea that the AIDS virus is a man-made entity produced by military researchers in their quest for a biological weapon was based, he said, on a Moscow radio allegation that the AIDS virus was the result of secret experiments carried out by the CIA and the Pentagon.[19]

The electronic media and the press of the Communist world relayed and commented profusely on the declarations of Dr. Seale. The latter, criticized by dissident Russian geneticist Zhores Medvedev, reacted by claiming that he could not be sure if the virus had been created by American scientists or by biologists at the Ivanovsky Institute in Moscow.[20]

John Seale then published a series of articles (one in collaboration with Medvedev, others without the latter's signature) about the first AIDS cases in the Soviet Union and the possibility that HIV was the creation of Moscow biologists.[21] Soviet officials, he wrote, "have cast some suspicion upon themselves by repeatedly accusing the United States Government, in overseas propaganda and in their own internal media, of developing the virus in military laboratories, and then proceeding to classify as secret all scientific research and information on AIDS in the USSR."[22] The virus may have been created by artificial selection and not by genetic manipulation. "If it is possible," Seale said, "that the pathogenic primate lentiviruses could have evolved naturally from a common ancestor since 1948, it must also be possible for them have evolved by artificial selection from the same common ancestor starting even more recently."[23]

The fundamental premises of Seale's conclusion, for which he had initially been reprimanded by Medvedev, remained the same, but he had switched sides: "It is entirely probable that the AIDS epidemic was started in the United States deliberately, by a hostile power, in the mid-70s."[24] It must not be forgotten, he observed, that the capitalist world is the principal victim of this diabolical product and that its spread can be more easily restrained in closed totalitarian states.

No such doubt troubled Rolande Girard, author of a violent book, printed in Paris in 1987 and completely devoid of subtlety. She had interviewed the Segals in Berlin, John Seale in London, and Robert Strecker, a physician in California. For Girard, their statements were the proof that HIV was a "genetic montage" made by American virologists under the patronage of the CIA and the Pentagon by combining fragments of visna virus with fragments of either bovine leukemia virus or Gallo's HTLV–I. According to Girard, the P4 laboratory at Fort Detrick was not operational before September 1977 and therefore she believed that the new virus was fabricated "at the very earliest in the fourth quarter of 1977."[25]

During the Third International AIDS Conference, held in Washington, DC, in June 1987, an anonymous leaflet was distributed to the participants claiming that the AIDS virus is "a bacteriological weapon developed by the United States against homosexuals and black people." Presented as a synthesis of three hundred scientific and medical documents, it was an unbelievable web of lies and pseudo-scientific statements.

Many arguments prove the impossibility of these allegations. But suffice it here to mention only three: HIV existed naturally before 1977; at that time, no scientist in the world had the biotechnological knowledge to "create" such a virus; and HIV–1 could not have been derived by an induced mutation or a genetic recombination from the visna virus.

Gallo related how, during a scientific meeting at Erice in Sicily, he was asked to give a special lecture to a group that included about ten Russian physicists who were especially eager to hear about retroviruses.

> When I finished, they asked me what I thought of the claim that the AIDS virus had been man-made in the United States. My response was quick but conclusive, and their response to mine was joy and applause. One of them (I believe it was Professor Eugene Valikhov of Moscow) promised me he would bring the problem to the attention of key people who knew the new Soviet leader, Mikael Gorbachev. I do not know whether this man was responsible for what happened a few weeks later when an official Soviet pronouncement condemned the notion of a man-made virus in the United States or anywhere else.[26]

Indeed, representatives of scientific and medical institutions in the Soviet Union and East Germany rejected the unofficial accusations of the press. Thus, for example, Viktor Zhdanov (1914–1987), director of the Virology Institute in Moscow, stated clearly in 1985 that the AIDS virus was a naturally occurring germ that, in his opinion, had existed for thousands of years in Africa.[27] Valentin Pokrovski, president of the Soviet Academy of Medicine, told the reporter of the French newspaper *Le Monde* that "no Soviet researcher had ever spoken of the artificial production of this virus. Like all the scientists of my country, I believe that the virus has a natural origin."[28]

After the fall of the Communist regime in the Soviet Union, confidential revelations confirmed that the misinformation concerning the origin of AIDS, truly a "media virus," had been produced and disseminated by the KGB. In the summer of 1992, an interview granted to the *Nouvel Observateur* of Paris, Yuri Kobaladze, spokesperson for Yevgeny Primakov, the new head of the Russian state police, admitted the responsibility of the former Soviet agents in the AIDS affair: "We have recently learned that the ex-KGB had conducted a misinformation campaign concerning AIDS against the United States. In 1983, at the beginning of the epidemic, they made use of foreign relay journals and then the Soviet press to promote the idea that the virus was the result of manipulation in the laboratories of the Pentagon. This was in response to the American campaign that had implicated the ex-KGB in the assassination attempt on the Pope."[29] Other Russians sought to vindicate the deliberate media intoxication by presenting it as a response to the unjustified accusations made in the *Wall Street Journal* during the spring of 1984 that the Soviet Union was using genetic manipulation to prepare highly lethal microbes (notably, influenza virus recombined with the gene of cobra venom).

A good example of an unorthodox hypothesis, that includes modern horrors, can be found in the explanation of Ernest Sternglass, professor at the University of Pittsburgh. He argues that the emergence of AIDS is related to strontium–90, the fallout from French experimental atomic explosions in Africa. With no basis in scientific facts, these ideas seduce the imagination with their symbolic implications.[30]

During the Second International AIDS Conference of June 1986, in Paris, the Russian virologist Viktor Zhdanov said in response to a journalist who had asked him if the Americans had fabricated HIV: "This is a ridiculous question. Perhaps it was the Martians."[31] Why not, indeed? Perhaps not the Martians of science fiction novels, but could not the virus have come from outer space? Surprisingly enough, the notion of an extraterrestrial origin for some "new" pathogenic germs had been clearly formulated and maintained as early as the fifth century BC by the Greek philosopher Democritus (c. 460–370 BC) and, more surprisingly, again recently by the famous astronomer Fred Hoyle (1915–2001) and even by Nobel laureate Francis Crick (1916–2004). Our surprise stems from the fact that Democritus did not know enough and Hoyle and Crick knew too much to advance such a hypothesis, which may in fact be logically possible but is technically unverifiable. Its principal failing is methodological: It does not solve the problem but pushes it back in time and space.[32]

Remarks on Currently Fashionable Hypotheses

Today's pandemic is the result of the superposition of at least two different epidemics, provoked by two distinct viruses: HIV–1 and HIV–2. They are genetically related, but their gene sequences are such that one could not be a descendent of the other. The HIV–2 epidemic might have passed unnoticed if the serious nature of the first had not sharpened the gaze of doctors and directed the research of virologists. This second epidemic was at first localized to a single area in West Africa, while the world dissemination of HIV–1 spread from three points: one in central Africa and two others on the shores of North America. The two American foci no doubt have a common origin, but it is not known if the outbreaks on the two continents are independent and somehow parallel or if one ignited the other. Clinical descriptions and analysis of frozen sera have demonstrated the existence of sporadic cases of AIDS since the middle of the century, in both the United States and the Old World.

During the past few years a considerable number of retroviral parasites of mammals, especially apes, have been isolated and found to be close to HIV. The genetic links between these infectious agents are now solidly established. From 1988 to 1990, molecular analysis of their genomes provided the basis for the construction of phylogenetic trees indicating the order of their evolution.[33] This order corresponds roughly to the evolution of their hosts. The genetic similarities between human and simian immunodeficiency viruses (HIV and SIV) indicate a common origin: A single ancestor is at the root of four branches of SIV/HIV (SIV of mandrills, SIV of green monkeys, SIV/HIV–2 of mangabeys, macaques, and humans, and the SIV/HIV–1 of humans and chimpanzees). It must be emphasized that, judging by the gene sequences, strains of HIV–2 are more distantly related to those of HIV–1 than are the strains of each of these groups to certain simian strains. The common origin does not necessarily mean that HIVs descended from recent SIVs. The relationship can be explained just as well by the passage of the simian viral strain to humans through parallel development from common ancestors.

The discovery of SIVs immediately suggested the hypothesis (formulated and defended especially by Myron Essex and Phyllis Kanki)[34] of recent cross-transmission with adaptation of simian strains to humans. Indeed, this hypothesis accounts reasonably well for most of the known facts concerning the species HIV–2 and its genetic links to the African mangabey virus,[35] although the discovery of the strain HIV–2 (D205), isolated from the blood of a native of Ghana, shows that things are not as simple as they appeared at the outset. This strain of HIV–2 is genetically closer to mangabey virus than to the common HIV–2.[36]

As for HIV–1, its origin via a recent transmission between apes and humans is possible. Most virologists consider this explanation to be probable but far from being proved. They are still seeking a missing link because the genetic distance between the common strains of HIV–1 and the supposed ancestor, SIV of green monkeys, is too great. In a few instances the gap appears to be diminishing: a strain close to HIV–1(HIV$_{cpz}$) was isolated from the blood of a Gabon chimpanzee in 1989;[37] while a human strain (ANT$_{70}$), which is closer to SIV of chimpanzees than to other human strains, has been found in 1990.[38] But the possibility that the HIV$_{cpz}$ resulted from an infection of an ape by a human cannot be excluded.

An ancestor of HIV–1 may have been a parasite of humans for a very long time, on the condition that it either remained confined in a few restricted populations or simply did not provoke serious pathological states. As Luc Montagnier (b. 1932) said in a recent interview:

> The classic hypothesis is that the virus came from Africa by transmission from primates to humans. But one can image that HIV–1 was endemic in a human population before the epidemic appeared, present for millions of years but not pathogenic. I am speaking of the HIV–1, because HIV–2 is very close to the simian virus, and it is very probable that it was transmitted from humans to apes. For HIV–1 the fact that different variants are found in different regions of the world—this diversity not being explained solely by human displacement—is in favor of the hypothesis of an endemic but non-aggressive virus. But a change was necessary before it became pathogenic for humans.[39]

The virulent HIV–1, cause of the present pandemic, is certainly the result of a new biological event—not necessarily a short-term event, the big bang of AIDS situated in the middle of this century—but possibly a process evolving over time that reached a critical threshold during the 1970s.

To explain the emergence of such a highly pathogenic virus, both biological and sociological explanations are invoked. An accidental mutation specific to the HIV ancestor may have occurred, or, as some say, a particular genetic recombination, which is in this case a sort of verbal solution, lacking hard data. According to Robert Gallo, the epidemic began with the passage of the simian virus to humans in the usual living conditions of Africa, where apes are hunted, slaughtered, and eaten. But it remains to be explained why the transmission happened at this point in time. Assertions that have been increasing in the press during the past two years suggest that the cross-transmission was due to medical experiments on African populations (specifically, trials on vaccines against malaria or polio). This accusation against Western medical practices is within reason but founded only on shaky circumstantial evidence.[40]

Another biological explanation appeals not to cross-transmission of viruses but to encounters of different microbes within the human species: According to Luc Montagnier, the present epidemic may have resulted from an alliance of the slightly virulent African HIV with American mycoplasmas selected by the use of antibiotics.[41] This clever hypothesis has the merit of promoting a political reconciliation between America and Africa, both "accountable" for the emergence of AIDS.

Sociological and Biological Selection Hypotheses

It is difficult for me to believe in the recent origin of the HIV–1 and HIV–2 viruses by unique and somehow parallel mutations or recombinations during a brief lapse of time. Our present understanding of the animal retroviruses makes it more likely to have been at a slower genesis, involving successive selections in an extremely variable viral pool. I am therefore convinced that it is possible, even probable, that there was never either a special animal reservoir for HIV–1 or a human reservoir in the form of an isolated and largely infected population. Perhaps for some centuries the virus existed throughout the world, scattered and manifest only at low level, in sporadic cases and mini-epidemics invisible to medicine before 1980. In the past, such a virus would have been less virulent and the routes for infection less wide open.[42]

Studies on the genealogy of retroviruses are clarifying the biological conditions without which the present pandemic would never have taken place, but they do not respond to the fundamental question: Why now? The answer goes well beyond the domain of virology. To resolve the problem of the origin of the AIDS pandemic, a complex of factors in which the biological and social are inextricably intertwined must be taken into consideration.

In 1978, for the first time in history, the conceptual and technical means were available for identifying and isolating a human retrovirus. And it was just then that AIDS began to spread. Another coincidence was soon added: the eradication, announced in 1977, of smallpox. It is hard to accept that these events could be totally independent, that their congruence was pure chance, but it is even more difficult to find any direct causal link between them. The insight came to me in the summer of 1987, during a symposium organized by Charles Mérieux (1907–2001) at Annecy: All these events were not causally related among themselves but instead flowed from a common source—that is, from the technological progress.

My explanation is both sociological and biological. The sociological aspect is not particularly original. It is obvious that the emergence of the present epidemic is linked to certain social changes that characterized the second half of the twentieth century: the intermingling of populations, a new type of sexual activity (liberalization of social mores, organized homosexual promiscuity), the massive use of IV drugs, changes in the techniques of blood transfusion, etc. As for Africa in particular, suffice it to recall the social upheaval and wars that followed decolonization, uncontrolled urbanization, the development of prostitution, the perverse side effects of modern medical treatments unsuited for this milieu (poor use of syringes, contaminated vaccines, etc.).[43] Invoking these social factors in an explanation of the beginnings of AIDS is perfectly legitimate but not sufficient. Their role cannot be understood without consideration of the biological aspect.

My biological hypothesis is original and unorthodox. It contains two principle ideas: the first concerns the biological properties of HIV and the laws of Darwinian selection; the second consists of an application of the general concept of pathocenosis to the specific case of AIDS.

One of the biological peculiarities of retroviruses is their great genetic variability. RNA viruses evolve more rapidly than do DNA viruses.[44] They possess an immense ability to adapt, because retroviral infection corresponds to the most elementary biological realization of a hypercycle (concept defined by Manfred Eigen): The rate of genome replication is influenced not only by the activity of the genetic model, which determines structure, but also by the concentration of a particular enzymatic subunit, encoded in the genome and thus dependent on the quantity of nuclear filaments. A very efficient feedback cycle is imposed on the cycle of replication. Natural selection exerts strong pressure on this type of "replicator."[45]

The AIDS virus is therefore extremely variable.[46] It evolves at a rate about one million times as great as that of eukaryotic DNA.[47] It produces non-pathogenic strains, defective strains, and highly virulent strains. Even now, in the midst of the expanding AIDS epidemic, some HIV strains are innocuous. Thus, for example, five seropositive persons, infected in Australia in 1982 by blood from the same donor, are still free of symptoms (at least in 1992).[48] At the other extreme of this spectrum of virulence, a strain of HIV, recently isolated in Zaire (HIV_{NDK}), greatly surpasses the pathogenicity of the reference strain (isolate of Montagnier, 1983).[49]

Highly pathogenic strains arise continually by random genetic drift. Nevertheless, if the routes for transmission are reduced, these strains are eliminated, since patients die before being able to infect at least an equal number of others. For a strain to be successful a patient must contaminate, on the average, more than one other person. This cannot happen if transmission is limited by biological conditions and restricted to sexual activity in the circumstances of traditional social life. Increase in pathogenic power is unfavorable for a parasite: Exhausting rapidly and killing the host diminishes the chances of effective transmission to other hosts. Conventional wisdom suggests that the direct action of evolutionary change will be toward reduced pathogenicity.[50] However, a handicap in virulence could be compensated by its association with a greater transmissibility. A set of new theories pertaining to the population biology of host-pathogen interactions suggests, in fact, that the outcome of evolution may be toward reduced or increased virulence, depending on the trade-off between pathogenicity and transmissibility.[51]

Obviously, transmissibility (or infectivity) is a property favored by Darwinian selection. The direction of selective pressure changes as a function of the number of effectively exposed individuals. When the number of people susceptible to infection is high, the rate of expansion of the virus increases with further increase in transmissibility even when the lethality is great. However, the increase of transmissibility (linked with the increase of virulence) works decisively against viral expansion in a population with a small number of exposed individuals. From the germ's perspective, the strategy that assures the best survival depends on the probability of infecting new hosts. Aggressivity is indicated when opportunities to infect new victims are frequent; peaceful coexistence is necessary if people not yet infected are for the most part resistant or if the routes of transmission are limited.[52]

These considerations bring us to an especially important fact. There is a critical threshold at which the biological situation is reversed catastrophically (in the mathematical sense): Below this threshold, selection maintains a weak level of virulence; above, it ensures the success of highly pathogenic strains of virus.

In the case of AIDS, the reversal appears because of the coincidence of several series of factors, which facilitate infection, all linked to postwar civilization. From the 1950s on, blood transfusion become increasingly frequent and an almost industrial manner of transfusion was developed, with the pooling of plasma from numerous donors, which notably increased the possibility of viral infection. Some forms of treatment for blood diseases opened up new routes for viral infection. Moreover, humans may have used drugs in the past, but intravenous drug use is quite recent. Last but not least, homosexual promiscuity within transcontinental networks functioned as a veritable culture medium for the AIDS virus. The historical novelty certainly does not reside in homosexuality itself but in the extent and degree of promiscuity. AIDS is not the only viral infection to take on epidemic proportions among American homosexuals. In Africa, other sociocultural factors, such as urbanization, alterations in heterosexual relations, and the allograft of Occidental medicine, produced similar effects.

As soon as the critical threshold is crossed, several poorly understood mechanisms can contribute to the random genetic drift to stimulate the appearance and dissemination of virulent germs: microbial synergism, activation of defective strains by genetic recombination, transmission to hosts in populations that are immunologically "virgin," etc. The observations of Paul Jolicoeur and other Canadians concerning the role of defective forms in the pathogenesis of a mouse disease analogous to AIDS suggest a similar role in humans.[53] Elsewhere, Patricia Fultz has shown that the virulence of mangabey SIV for a host, with which it had been in good biological equilibrium, can be artificially increased by successive inoculations of macaques with a strain of this virus.[54]

Already Louis Pasteur (1822–1895) had imagined that an increase in a microbe's virulence could result from an increase in the frequency of passing through sensitive hosts and that this might explain not only traditional epidemic diseases but also the appearance of new diseases:

> What is an inoffensive microscopic organism for humans or for a certain animal? It is a being that can develop in our body or in the body of this animal; but if this microscopic being managed to penetrate another of the thousands and thousands of species in creation, nothing proves that it could not invade it and make it sick. Its virulence, reinforced by successive passages through individuals of this species, could reach a level at which it could kill one or another large animal, humans or certain domestic animals. By this method new virulences and new contagions can be created. I am inclined to believe that this is how smallpox, syphilis, plague, yellow fever appeared across the ages, and similarly that phenomena of this type explain how certain great epidemics, such as typhus for example, have appeared at one time or another.[55]

It is not impossible that occasional exchanges of virus between apes and humans have taken place in the past—transmissions that could give rise to virulent strains yet remain free of serious epidemiological consequences. The decisive factor was not the appearance of virulent strains but their statistical probability of being transmitted within a population. In the absence of combustible material, even a highly pathogenic virus is only a spark, which may of course consume the infected individual but cannot provoke a blaze.

I compared the process that gave rise to epidemic AIDS to that which favors the growth of a cancer by the rupture of equilibrium between immune defense and the inevitable and incessant cellular errors, or touches off an atom bomb by exceeding a critical mass of fissionable material. The American press came to this idea rather late but with quite an uproar. The cover of *Newsweek* on March 22, 1993, proclaimed: "Taming the AIDS virus. New research suggests HIV is an old virus that turned deadly."[56] In the text, signed by Gerard Cowley, the "new view" is credited to Paul Ewald, an evolutionary biologist at Amherst College.[57]

My explanation includes one other necessary and complementarity idea: a significant factor that changed the direction of selective pressure is the marked decrease in other infectious diseases. In 1969, I defined and named the concept of pathocenosis according to which the frequencies of diseases obey certain rules and can be studied using mathematical models.[58] The frequency and overall distribution of each disease depends a great deal on the frequency and distribution of other diseases in the same population. The history of any disease is thus a tributary of the history of all diseases. The limits of space prevent me from entering into detail about the definition and historical dynamics of pathocenosis, but suffice it to say that its application in the case of AIDS is particularly instructive and that, faced with what I have seen since the beginning of this epidemic, I would have invented this concept now, if I had not already invented it.

Medical progress contributed to the new epidemic as much by rupturing the fragile pathocenosis—that is, by suppressing competing diseases—as it did by facilitating the transmission of the infection, notably through new means of direct blood contact. The most important change in morbidity in the twentieth century is the spectacular reduction in tuberculosis. At the beginning of this century, pulmonary phthisis and other forms of tuberculous infection were still widespread and deadly, although generalized improvement in living conditions had contributed to a decline in the previous century. Until the 1950s, mortality due to tuberculosis decreased greatly: Medicine had transformed the lethal forms into chronic forms. From then on, vaccination and chemotherapy lowered the specific morbidity.

In the pathocenosis of the first half of our century, tuberculosis and a few other diseases made sporadic cases of AIDS invisible and prevented the dissemination of this disease. Not merely a screen behind which the misdeeds of HIV precursors could hide, these diseases constituted a real barrier. Virulent strains of HIV were rapidly eliminated or circumscribed when they were introduced into a population greatly infected by tuberculosis and other so-called opportunistic diseases.

In our explanation of epidemiological changes, we should not neglect the biological links between infectious diseases. A very illuminating example is the correlation between tuberculosis and leprosy. Intense study of the genetic and biochemical properties of HIV has shown some remarkable facts, the biological significance of which is unknown at the present time. Thus, for example, André Carson's team at Lille discovered that an antigen of HIV–1, the small regulatory protein called Virion Infectivity Factor (vif), corresponds exactly to one of the surface antigens of the helminth, *Schistosoma mansoni*, the parasitic worm responsible for African bilharziosis. A partial immunological cross-reactivity can be demonstrated in the laboratory between AIDS and bilharziosis, but the natural impact of this phenomenon is unknown.[59] It is worthy of note that bilharziosis is hyperendemic in Zaire and Burundi, precisely the countries that were the initial site of the African AIDS epidemic.

In conclusion, changes in social behavior have contributed to the appearance of AIDS, but they alone did not suffice for a crossing of the critical threshold of a biological process that shapes HIV virulence. Technological progress must also be blamed; the shocking paradox is that it is even more responsible. Blood transfusion is a medical success but also a vehicle of the virus. Elimination and control of diverse infectious diseases were great victories, but they opened the door for AIDS. Thus, the situation is a corollary of what Edward Tenner (b. 1944) calls the revenge theory: "Technological progress has changed our world, but the world seems bent on getting even, twisting our cleverness against us."[60]

PART TWO

Experiments and Concepts in Life Sciences

First Steps in Claude Bernard's Discovery
of the Glycogenic Function of the Liver

The Archives of the Collège de France in Paris are in possession of a very large and impressive collection of the notebooks, laboratory journals, and other scientific manuscripts of Claude Bernard (1813–1878). These papers are now organized and available for scientific research.[1] Some notebooks and papers give significant documentary information on Bernard's philosophical background and his position between the materialistic doctrine and the vitalistic conception of life.[2] For the historian of science, however, more interesting perhaps are Bernard's laboratory journals and day-by-day reflections on physiological problems.[3] In his famous *Introduction*[4] he accords to his own discoveries the dignity of paradigms. Thus, a detailed study of all the steps in his creative activity is a necessary condition for acceptance of his findings as epistemological examples. An analysis of his laboratory journals reveals in many cases an important historical inconsistency. On the one hand, his original manuscripts suggest a very complicated gradual development of his discoveries, while on the other hand, his published works show a tendency toward a secondary rationalization, that is, a very strong *post hoc* simplification of facts. If the examples quoted in his *Introduction* are all logically consistent, many of them are chronologically incorrect and simplified to the point that some very important steps are masked.[5] I will try to illustrate this point by an example that seems minor at first glance but that actually is of extremely great importance in Bernard's research work. The unexpected result of one experiment changed the whole direction of his investigation of the destination of sugar in animal organisms.

When he began his experiments with sugar, Bernard shared the view of Dumas and Boussingault that it was formed by green plants, introduced in animals by alimentation, and destroyed in them by a special process of combustion.[6] Animals were supposedly able only to break down sugar supplied by vegetables. Bernard accepted Liebig's opinion that sugar was the fuel of life, and he believed that the action of combustion took place either in the lungs (Lavoisier's initial hypothesis) or in the general capillaries (hypothesis of Lagrange and Hassenfratz). In 1843 Bernard discovered that an animal organism could directly utilize only sugars of the "second species" (for example, grape sugar) and that sugars of the "first species" (cane sugar), when injected into the blood of animals even in very weak doses, passed into the urine. He noted, too, that gastric juice could transform cane sugar into a form capable of assimilation—that is, of destruction in the animal organism.[7]

The next step in Bernard's work was later summarized by him in the following way:

> Thereupon I wished to learn in what organ the nutritive sugar disappeared, and I conceived the hypothesis that sugar introduced into the blood through nutrition might be destroyed in the lungs or in the general capillaries. The theory, indeed, which then prevailed and which was naturally my proper starting point, assumed that the sugar was destroyed in animal organisms by the phenomena of combustion, i.e., of respiration. Thus sugar had gained the name of *respiratory nutriment*. But I was immediately led to see that the theory about the origin of sugar in animals, which served me as a starting point, was false. As a result of the experiment which I shall describe further on, I was not indeed led to find an organ for destroying sugar, but, on the contrary, I discovered an organ for making it, and I found that all animal blood contains sugar even when they do not eat it. So I noted a new fact, unforeseen in theory, which men had not noticed, doubtless because they were under the influence of contrary theories which they had too confidently accepted. I therefore abandoned my hypothesis on the spot, so as to pursue the unexpected result which has since become the fertile origin of a new path for investigation and a mine of discoveries that is not yet exhausted.[8]

This famous text emphasizes an invaluable general recommendation. But in some details it seems quite vague, even to the point of obscurity. With his sentence "But I was immediately led to see . . ." Bernard gives the impression, and wrongly so, that he changed his mind very quickly after the beginning of his experiments on sugar destruction in an animal organism. And what is more important, he does not really explain why—at which concrete occasion—he abandons the prevalent theory. Curiously enough, he never really elucidated this point,[9] and in his fundamental publication on the discovery of the glycogenic function of the liver, he presented his experiments without chronological order, without dates, and following a logical development completely independent of the historical linkage of his experiments and the real evolution of his thought.[10]

He was not "immediately led to see that the theory about the origin of sugar in animals . . . was false," because he started his experiments in 1843, increased their number and perfected them from 1844 to 1847, and finally understood that he was on the wrong track in August 1848. His notebooks contain descriptions of large numbers of experiments concerning the search for the location and mode of destruction of carbohydrates after ingestion, intravenous injection, or other introduction into an animal organism.[11] These experiments have never been published because Bernard was fully aware that they brought forward nothing new and represented only a failure. The positive side of all of this lengthy previous work is that Bernard elaborated—with his friend, the young chemist Charles-Louis Barreswil (1817–1870)—on the chemical testing of sugar, further that he understood better the first phases of the digestion of starch, and that he observed the influence of the nervous system on the presence of sugar in blood and urine. Certainly, by his long and numerous experiments he became sensitive to all the possible physiological implications of the presence or absence of sugar in various parts of the animal circulatory system.

Beginning in 1845 Bernard became interested in the clinical problems of diabetes. He observed patients[12] and formulated his first theory of the pathogenesis of this disease.[13] According to Bernard's first opinion, diabetes is "a nervous affection of the lungs." For the modern reader this theory is very surprising, but actually it was a very logical conclusion from these four premises: 1) sugar cannot be synthetized in the animal body; 2) it is normally destroyed in the lungs; 3) the principal symptom of diabetes is the presence of undestroyed sugar in the urine; and 4) the nervous system controls the breakdown of sugar in the lungs. Claude Bernard discovered that after cutting the pneumogastric nerves in rabbits the pulmonary functions are affected and glucose passes undestroyed into the urine.[14]

One important problem was whether or not the blood of diabetic patients actually contained sugar. Thomas Willis (1621–1675) was the first to believe it.[15] In the eighteenth century, Dobson, Cawley, and Rollo tried to extract sugar or sugar-like substances from the serum of diabetic patients. They wished to demonstrate that glycosuria is merely a sequence in glycemia. But none of these attempts produced any definite conclusion. The sweet taste of blood was not sufficient proof, and chemical analysis by alcoholic fermentation gave generally negative results. Thus P. F. Nicolas and V. Gueudeville, Soubeiran, Vauquelin, and other authorities on this subject at the beginning of the nineteenth century were not ready to accept the theory of diabetic glycemia. Their negative results can probably be explained by the fact that the analyses were performed on old blood, after glycolysis. In 1835 an Italian chemist, Felice Ambrosioni (1790–1843), was the first to give definite proof of the presence of sugar in the blood of a diabetic person. His demonstration was based on the alcoholic fermentation by yeast of blood sugar.[16]

Even before it was definitely proved that sugar could be found in the blood of persons with diabetes, it was known that this substance could be present in the blood of healthy animals, at least in some animals under special conditions. In 1826, F. Tiedemann and L. Gmelin demonstrated the presence of fermentable glucose in the intestines and venous blood of healthy dogs after ingestion of starch.[17] In England, MacGregor[18] confirmed the observation of Ambrosioni, and Thomson,[19] a chemist of Glasgow, found that chicken blood normally contained a certain amount of sugar.[20]

In France, François Magendie (1783–1855) discovered, independently of the aforementioned authors, that sugar can be found in the blood of normal rabbits and dogs after they had been fed on starch or potatoes.[21] After Magendie's experiments, performed during his lectures at the Collège de France in 1846, the majority of physiologists and physicians agreed in supposing an alimentary origin of sugar and considering glycemia as a physiological phenomenon compatible with health but inconstant—being a result of ingestion of special kinds of food.

Thus, the presence of sugar in blood was considered to be either a pathological or an accidental fact. It was Bernard who discovered that glycemia was a normal and constant phenomenon, largely independent of alimentation.[22]

One unpublished manuscript permits us to have a real understanding of how this discovery occurred: This is Bernard's laboratory journal *Ms. 7c*, compiled from 1846 to 1848. The beginning experiments are of no interest, because, having taken a wrong turn, Bernard was unable to progress. Until May 1848 he attempted to answer a badly formulated question. Yet at this time he believed that sugar must be destroyed somewhere within the organism. His main attention was evidently directed toward the lungs, and in the last week of May he observed what happened when grape sugar was exposed in vitro to the pulmonary tissue of freshly killed animals. After ten to twelve hours the sugar disappeared, and Bernard concluded that the lungs contained a special ferment for the destruction of glucose.[23] But following faithfully his method of experimental research, he proceeded to a counterproof. This he did by mixing sugar with tissues of liver and other organs. He obtained positive results, and the problem became even more obscure than before he started his experiments.

On the last day of May, he injected one gram of grape sugar into the jugular vein of a dog, extracting at the same time blood from the carotid artery. This blood contained a large amount of sugar. The conclusion was evident: Glucose is not destroyed in the lungs, because blood must pass by these organs in order to move from the jugular vein to the carotid artery.[24] Bernard guessed that perhaps the combustion of carbohydrates took place in the general capillaries or in the liver.[25] Numerous experiments, carefully executed by Bernard during June and July of 1848, were strongly opposed to the theory of pulmonary combustion. Grape sugar was injected into the jugular veins, or starch introduced into the stomachs, of rabbits and dogs. Then either blood was taken from various parts of the animals or they were killed and blood extracted separately from different organs. Sugar was present in all samples. Bernard was unable to find a rule for its quantitative distribution.

His friend Quevenne, pharmacist at the Hôpital de la Charité, extracted and purified blood sugar from a diabetic patient, and Bernard showed that in physiological experiments there is no difference between grape sugar and the sugar of diabetics.[26]

In July 1848, Claude Bernard discovered some important facts: 1) the transformation of cane sugar into grape sugar by the action of gastric juice is not performed by gastric acid but by some special "organic matter";[27] and 2) sugar is always present in the vitreous humor of the eye of a dog and also in the white of a chicken egg.[28] This last finding was not immediately published, and when in March 1849 an Irish physician named Aldridge[29] discovered this fact independently of Bernard, he claimed priority.[30] Preserved notebooks provide proof that Bernard's claim was well founded. This discovery is not without historical significance, because it invalidated the older demonstration of the presence of sugar in the blood of diabetic persons. Actually, Ambrosioni and MacGregor added egg white to blood before testing its sugar content.

Bernard observed that Barreswil's copper reagent did not react well with sugar in the presence of fibrin. He imagined a new theory of the pathogenesis of diabetes. Thus, sugar was supposed to be destroyed in blood, fibrin having some important function in this destruction. And diabetes was nothing more than a stopping of this destructive process probably by some chemical disorders involved in the synthesis and distribution of fibrin.[31]

This original point of view gave Bernard the possibility of foreseeing new experiments. Analyzing the blood in the vessels before and after each single organ, Bernard wished to eliminate, step by step, the ancient theory that sugar combustion is located in a particular area of the organism. The first experiments seemed to confirm his new working hypothesis. In dogs fed on a carbohydrate-rich diet, the blood from the hepatic veins and vena cava contained sugar; thus, it was not destroyed in the liver. Sugar was also present in both ventricles of the heart, meaning that it had not been destroyed by the lungs.

From many laboratory notes it is clear that Bernard attained these results and was very happy that they corresponded to prediction. But as in all cases, he wished to assure the results by counterproofs. One dog was submitted to a noncarbohydrate diet then killed by section of the spinal bulb. Blood was taken from the portal vein, from both ventricles, and from a peripheral artery. The results were completely unexpected, astonishing, and puzzling. The blood of the portal system contained enormous quantities of sugar; the blood from the heart contained sugar but in a small amount; and the arterial blood showed only traces. Chyle had no sugar whatsoever.

It is perhaps useful to publish the exact text of Bernard's laboratory notes concerning this crucial experiment:

32 Dauphine Street. August 1848. Experiments on sugar in blood. Peristaltic movements. Blood is drawn from the jugular of a dog that fasted for six days and was only given water; serum of fresh blood, right after its coagulation, gives traces of reduction when the blue liquid is used. The next day it doesn't anymore. Part of the same serum is treated with alcohol and then evaporates, then is mixed again with water and beer yeast and produces only a few bubbles of gas; fermentation, thus, is barely noticeable. One must admit that there was very little serum being used, around fifteen grams, and the quantity of sugar was also minimal.

On the same dog, under other circumstances, when it fasted for only two days, freshly obtained serum also gives reduction with copper tartrate.

The same dog, now back on a meat diet and which, during eight days in a row only had raw bits of meat from the butcher, is killed by severing the medulla to collect the blood. Respiratory movements came to a complete rest, as usual, once the medulla was cut. The eye on one side remained sensitive but not on the other. I opened the paunch immediately, while the heart was still beating. Chyliferous vessels[32] were full of white chyle[33] and the intestines, contained in the belly as the animal laid on its back, did not show any sign of peristaltic movement. As I pressed the aorta in the chest with two fingers, the peristaltic contractions started intensely and were unstoppable. Those movements, it seems to me, became slightly weaker when I released the pressure, but they increased when I pressed my fingers again. All this did not last long, though, because the heart's movement soon came to a halt. What was salient was how peristaltic contractions started when I pressed the thoracic aorta.

Extraction of blood. I drew separately: 1° blood from the portal vein upon its entrance in the liver; 2° blood in right and left ventricles of the heart; and 3° blood from the neck wound caused by the severing of the medulla.

Those three samples of blood were left alone to coagulate. A few moments later, they all had coagulated, presenting a lactescent whitish serum. (Venous blood had the same outlook; did chyle spill over in it or was it venous blood returning?) White chyle extracted from the thoracic canal coagulated after a few moments.

I tried copper tartrate on the three fresh serums and with the chyle, which was also fresh. 1° When treated directly, the chyle from the portal vein gave reduction in huge quantity. When precipitated with dry sodium sulfate so as to get a colorless liqueur, reduction also obtained profusely.

2° Heart chyle treated directly with copper tartrate gives a very clean reduction although less abundantly than blood from the portal vein. 3° Serum from the neck blood treated directly with copper tartrate gives hardly any reduction at all. After being treated with sodium sulfate the reduction is still equivocal. 4° Chyle serum treated directly with copper tartrate gives no discernible reduction.

How is it possible to find sugar (or a reducing material) in blood of the portal vein? I looked into the intestine: 1° Reduction of copper tartrate did not obtain from the liquid of the stomach that contained meat about to be digested. 2° Bilious intestinal liquid failed to reduce copper salt completely. 3° Urine treated beforehand with sodium sulfate also did not reduce salt of copper.

Heart serum kept until the next day still had some sugar, which means it reduced tartrate. This experiment is exceedingly strange. From it one can comprehend nothing. Sugar is apparently formed in the portal vein. By what organ, by what mechanism?

It will be necessary to take this blood from the portal vein of a fasting dog and see if one will find there that material that reduces. If sugar is formed from an alimentation other than that of starch, the question of diabetics is singularly complicated. It will be necessary to see if that reducing material (sugar or otherwise) disappears quite rapidly, for the blood of the heart contained less of it and the blood from the neck only in a very equivocal fashion.

What, therefore, is the organ which would form that sugar or that reducing material?[34]

Where did the sugar come from in an animal without an alimentary supply of carbohydrates? Bernard wrote in his journal, as if crying in surprise: "It is absolutely incomprehensible!"

This experiment was done in the laboratory of Theophile-Jules Pelouze (1808–1867), a famous chemist and Bernard's mentor. The exact date of the experiment is not stated, but its location in the laboratory journal places it between the 10th and the 17th of August 1848, probably closer to the latter date. It is significant that in this case Bernard conducted at the same time and on the same animal two different experiments, one concerning the regulation of conditions of peristaltic movements in the intestines and the other concerning the metabolism of sugar. It is clear enough that the second part was considered only as a routine counterproof of previous experiments.

Bernard's notes express his astonishment. Actually, the discovered facts completely contradicted his working hypothesis and the generally accepted ideas on animal physiology. The presence of sugar in the blood of an animal without an alimentary supply of carbohydrates was such an incredible finding that Bernard, as we see from his journal, doubted the specificity of the copper reagent.[35] This "reductive matter"—was it really sugar? He decided to repeat the experiment, using other methods of chemical analysis.

Bernard understood immediately that the presence of sugar in the portal vein of a dog without sugar in the chyle had far-reaching consequences and that it would completely change existing theories of the pathogenesis of diabetes. He decided to repeat the experiment on a starving animal, and he posed the crucial questions: 1° where and by what mechanism was sugar formed in animals? 2° which animal organ performed this "vegetable" function?

Within a few days the basic fact—namely, the presence of nonalimentary sugar in mammalian blood—was confirmed by new experiments. On August 21 Bernard obtained positive evidence of the presence of glucose in the blood of a dog fed on lard and tripe exclusively. Thus he no longer hesitated to affirm that "there is a formation of sugar at the expense of fat."[36] But the great innovation of this experiment lies in the results of chemical examination of tissues taken from various abdominal organs. I cannot resist the temptation to quote the findings in Bernard's own words: "I collected tissues 1° from the spleen; 2° from mesenteric lymph nodes; and 3° from the liver, and I searched for sugar. There was almost no sugar either in the spleen or in the lymphatic ganglia although the substance appeared in *enormous* quantity in the liver."[37]

What a surprising result! Many questions assailed Bernard's mind. Was the presence of sugar in liver tissues a physiological phenomenon? Was it exclu-

sive to the liver or was it a property shared by other organs? Was it a characteristic of dog's liver only or was it common to all animals? After a few days of
feverish research, Bernard had found the answers to all these questions. Some
extracts from his laboratory journal give good evidence of his investigations:

> August 22nd 1848. I obtained veal and beef liver from the butcher. Using the
> reactive from fermentation I found enormous quantity of sugar in both.
>
> August 23rd 1848. At the Hôpital de la Charité I took three pieces of liver.
> 1°A grainy piece from an old man who died emaciated of I don't know what; I
> found no sugar using the reactive. 2° A very soft piece from a fatty woman who
> died of I don't know what; sugar was not found using the reactive. 3° A piece
> from a apparently healthy man poisoned with arsenic acid to whom iron perox-
> ide was administrated. I found enormous amont of sugar using the reactive and
> fermentation.
>
> August 24th. Liver from a dog used in the experiment recounted on page 387,
> dead since three days, still contains enormous amount of sugar after treated with
> reactive. It was necessary to place the liver in water and to work with this water,
> which had a strong, paludinous smell.
>
> August 25th. Liver from an albuminuric person[38] suffering from organic heart
> disease, a long and chronic illness; very infiltrated. The liver is congested; using
> tartrate, I could find traces of sugar.
>
> Traces of sugar are also found in the liver of a man who died from a heart
> disease. One will have to dose in different cases.[39]

Thus the presence of sugar in human and bovine liver was clearly demonstrated. On August 25 Bernard found sugar in the livers of a frog, a rabbit, a
capon, and two calf fetuses killed in the slaughterhouse at Popincourt, but he
was not able to find sugar in the liver of a ray or a lizard.

These phases of Bernard's work culminated in him presenting a note to the
Academy of Science on August 28, 1848. In this communication, signed by
Claude Bernard and Charles Barreswil as coauthors, it was stated that sugar
extracted from the liver is glucose by chemical nature, that it cannot be found
under physiological conditions in any other organ, and that its presence in the
liver "is a physiological fact which is completely independent of the nature of
alimentation."[40] Bernard and Barreswil presented as evidence to members of
the academy a sample of alcohol originating from the fermentation of liver
sugar.

The crucial experiment quoted in this paper corresponds to the first experiment of the second series in Bernard's classical memoir on the origin of
sugar in the animal body, presented on October 21, 1848, before the Société

de Biologie.[41] But there are significant differences between the original experiment and the published text.

In August 1848, Claude Bernard was able to demonstrate only the first of the four conclusions quoted in his October memorandum. He was sure that "in the physiologic state, there exists constantly and normally the sugar of diabetes in the blood of the heart and in the liver of man and animals," but he only conjectured without any experimental proof that "the formation of this sugar takes place in the liver."

Actually, Claude Bernard was immediately led to see that the old theory about the nutritive origin of animal sugar was false and that the liver produced sugar. But this "immediately" meant immediately after his crucial experiment in August 1848, and not immediately after the beginning of his research concerning the metabolism of sugar. Of course, it was easy to explain the presence of sugar in the blood of the portal vein in spite of the fact that the current of the blood should carry in an opposite direction all substances found in the tissue of the liver. Bernard supposed that the blood rich in sugar had seeped back into the portal vein when, by the opening of the abdomen, the pressure on the viscera ceased.

The real demonstration of the glycogenic function of the liver was accomplished during September and October 1848 with a series of experiments characterized by ligatures of the blood vessels of living animals and determination of the sugar in different parts of the circulatory system. Bernard showed that in properly performed experiments there is no sugar in the blood of the portal vein of an animal that is starved or fed on meat. His main argument in favor of the theory of the glycogenic function of the liver was precisely the absence of sugar in the portal vein and its presence in the suprahepatic veins and in the arterial blood.

How lucky he was to ignore some facts. First of all, there is in every case some amount of sugar in the blood of the portal vein. It was only by a special property of his chemical test that a gradual difference was transformed into an all-or-none reaction. The large amount of sugar in the blood of Bernard's dogs resulted from the manner of killing (section of the medulla oblongata) and should be interpreted as an exceptional, pathological condition. It is astonishing "how much instinctive judgment and even sheer luck contributed to a discovery which Bernard, with a good deal of justification, believed to be based upon the strictest experimental proof."[42] And how interesting it is to measure the extent to which a great scientist reconstructs his own previous thoughts to fit his later point of view.[43] The next steps in Bernard's work on the glycogenic function of the liver are, from this point of view, even more illuminating. In this case, as probably in the historical analysis of all other scientific discoveries, it is of invaluable help to resort to original first-hand documents. The importance of the systematic conservation of this kind of document, especially of laboratory journals, can hardly be overestimated.

FIVE

The Causes and the Nature of Ageing

By way of introduction, it may be useful to define the scope and object of gerontology and geriatrics.[1] Gerontology is the study of biological ageing processes, physical and psychosocial properties of aged organisms, and social problems of the aged. Geriatrics is a medical specialty dealing with the study, treatment, and prevention of diseases of the aged.

The term *geriatrics* was coined by the American physician Ignatz L. Nascher (1863–1944). He first used it in his programmatic article published in the *New York Medical Journal* on August 21, 1909.[2] The concept, however, is much older than the name invented by Nascher: In the past, geriatrics was called *gerocomica*, *gerocomics*, *gerontocomia*, or most often by some descriptive names in Latin or in national languages. Admittedly, the term *gerocomica* and its derivatives used by Galen related, for the most part, to a restricted field of geriatrics—to the hygiene of old age—but some writers, such as François Ranchin (1560–1641), included in it the treatment of diseases of the aged as well. The term *geriatrics* is well chosen. It is analogous to pediatrics, a branch of medicine concerned with the diseases of the first portion of human life, just as geriatrics is concerned with the diseases of the later part of human life.

There has been much disagreement among modern authors as to the definition of gerontology, especially with regard to the relationship between gerontology and geriatrics. Most European writers consider gerontology a wider scientific discipline, which embraces geriatrics as its clinical part. French and some German writers think it quite superfluous to introduce the neologism *geriatrics*, the term *gerontology* having, in their opinion, a more appropriate and broader sense. American authors, on the contrary, look upon gerontology and geriatrics as two quite different disciplines, partly overlapping. In their opinion, the relation between gerontology and geriatrics may be compared with that between psychology and psychiatry: psychology as a scientific discipline predominantly occupied with the study of healthy, normal processes, and psychiatry as a specialized practical branch of medicine dealing chiefly with pathological, morbid changes.

The objects of gerontologic studies are (1) the ageing of the organism (i.e., changes in the living systems with the passage of time) and (2) the age of the organism as an inevitable sequel of ageing processes. What indeed is ageing? Let me first set forth a formal definition as an introduction to a special chapter reviewing all the attempts made so far to come as near as possible to a real definition of ageing, to an explanation of the essence, the causes, and the primary symptoms of this process, which from the most ancient time has represented an unparalleled challenge to mankind and has been the object of most intensive studies, since it is something that irrevocably and mercilessly deprives living beings of their most valuable treasure: vitality, the vital force, the life's pleasure, and, at the end, life itself.

Ageing is a progressive and irreversible changing of the structures and functions of living systems. When defining ageing as progressive changing, we clearly state that it is a function of time. In vain is the entreaty of the great French poet Lamartine (1790–1869) in "The Lake" addressed to Time to halt its course and to throw its anchor but for a single moment into the sea of eternity, just as vain is man's desire to stop the ageing of the human or any other organism.[3] Changing our cells makes the existence of our organism possible, just as the continuation of species cannot be maintained without a constant succession of organisms. Ageing and death are necessary. Though harmful for the individual, they are useful and necessary for the continuation of the species and of life in general. The more often the exchange of generations of a species occurs, the more secure is its continuation and its biological adaptability. . . .[4]

The causes of ageing have not yet been scientifically explained, nor has there been any answer to the question of what is essential and primary in the process of senile involution. All the ideas recorded so far, including the most modern ones, are hypothetical and have a historico-medical rather than an actual scientific implication.[5] Ancient opinions about the causes of ageing can be divided into two groups: one supporting the idea that the cause of ageing is a gradual loss of something that is important for the maintenance of life (this "something" being explained by various authors in various ways—namely, as an energetic, psychic, or material factor); according to the other group, ageing is caused by the accumulation, the surplus of something that is deleterious to the organism (i.e., endogenous or exogenous intoxication of the organism). These two logically opposed concepts, more or less modified, have been maintained up to the present time. They have recently been supplemented by some more complex explanations. From a historical point of view, the hypothesis that ageing results from a gradual loss of specific vital energy is of the greatest importance.

The Causes and Nature of Ageing in Antiquity

The most ancient explicit scientific attempt to explain the cause of ageing may be found in Aristotle's work *On Youth and Old Age, Life and Death, and Respiration*. Aristotle (384–322 BC) thought that senescence was due to a gradual waste of "innate heat" (ἐμφυτόν θερμόν), which every living being possessed from the beginning of its individual life.[6] The center of "innate heat" is the heart. Blood vessels transport the heat throughout the body, which animates organs and limbs. This was not Aristotle's original assumption, however, but rather a systematized version of more ancient thoughts. A careful analysis of Hippocrates's works (460–375 BC), particularly his treatise *On Regimen*, shows that they entail the hypothesis that ageing is due to the loss of "innate heat." Traces of this concept can also be found in pre-Socratic philosophers such as Heraclitus of Ephesus (c. 535–475 BC) and Parmenides of Elea (c. 515–450 BC), and even in the Hebrew Old Testament. Translated into modern scientific language, the ancient idea of "innate heat" means—as L. Luciani rightly points out—the "sum total of the virtual energies and hereditary tendencies accumulated in the germ."[7] Hippocrates and Aristotle's opinion was based on a right observation—namely, that the production of heat in the aged is lower than in the young. To put it in modern terms: Senility is the result of a gradual diminution of metabolic processes.

Galen (129–217/216) goes a little further and connects observations on the lowering of temperature with clinical experience of dehydration of aged organisms. According to him, senescence is due to the loss of "innate heat" caused by a diminution of the moisture of the body.[8] Galen thinks that in old age the amount of blood is reduced, a process by which the chief fuel maintaining the flame of life is taken away.[9] This idea about the importance of the humors of the body dominated medicine up to the eighteenth century. Scholastics, for instance, compared life with an oil lamp that burns itself out.

The treatise on geriatrics written by Montpellier professor and court physician André Du Laurens (1558–1609) is a paradigm of the Galenic interpretation, which was accepted by physicians of Salerno, Arabian scientists, and professors of the first Western universities.[10] According to Du Laurens, senescence is due to a permanent struggle between the four humors and the "innate heat." Heat lives on moisture, he says, "just as oil maintains the flame in [an] oil lamp."[11] The moisture of the body is restored by food, but this process is never perfect. Thus, just as wine is diluted by water, our bodily organic substance is diluted by foodstuffs. The organism becomes dry and cold, and at the end the flame of life goes out.[12] In his famous 1860 lectures on the chemical history of a candle, Michael Faraday (1791–1867) very charmingly compared life with the flame and ageing of the organism with the burning of a candle. The most condensed formulation of Galenic views on senility, however, was given by Italian physician Aurelio Anselmi: "Old age . . . is a failure of life, which results from a lack of innate heat caused by a natural decrease of moisture."[13]

The Galenic hypothesis was worked out in detail by Italian physician Gabriele Zerbi (1445–1505), a Renaissance pioneer of geriatrics,[14] who says that in youth, food is evenly distributed throughout the whole body, a process that produces a harmonious supply of moisture and heat. By a continuous repetition of this process, the loss of moisture of the body gradually increases, bringing about rigidity in bodily tissues. While "innate heat" slowly eats up the moisture of the body and dries up the organism, moisture, in turn, steadily uses up "innate heat." Here, the cause of senescence is an inherent contradiction between body heat and body moisture. By the loss of fluids, the organism becomes rigid, the distribution of the foodstuffs more and more uneven, and, consequently, the relation between moisture and heat more and more upset. The cause of life and death is a continuous struggle between opposite principles.

Vitalist and Mechanist Theories of Ageing in the
Eighteenth, Nineteenth, and Twentieth Centuries

The eighteenth century was characterized by the adaptation of ancient and medieval concepts to a new scientific trend consisting in the creation of all sorts of speculative vitalist systems. There was a widespread belief at that time in a *vis vitalis*, a force governing all living phenomena, which was, according to the formula of the School of Vitalism, the subtlest natural agent, subtler than light or magnetism but fundamentally akin to these kinds of energy. "Naturphilosophers" and physicians tried to explain senescence by a generalized and superficial hypothesis of the weakening of this "vital force." In fact, their attempts were merely a simplified version of an ancient hypothesis where the term "innate heat" was replaced by "vital force." The typical advocates of this idea were the Montpellier physicians of the eighteenth century and the German Romantics of the early nineteenth century. Their most authoritative representative was Christoph Wilhelm Hufeland (1762–1836). Although he was a very clever scientist with a good critical sense, and an outstanding clinician and geriatrician, he supported a naïve hypothesis according to which senescence was due to a gradual loss of the "vital force" (or rather: *vital energy*) given in a certain measure to each organism at birth.[15] On the basis of similar assumptions, philosopher Herbert Spencer (1820–1903) formed the concept of a "capital of vital energy" given to every organism at the beginning of its individual career and consumed slowly by vital activity. English physician Erasmus Darwin (1731–1802), the uncle of the famous biologist, also advanced a hypothesis according to which senility was the result of the exhaustion of vital irritability. Erasmus Darwin's idea was given a modern form by Otto Lubarsch (1860–1933), a pathologist who thought that senility was only a peculiar sort of cell fatigue characterized by irreversibility.

In the nineteenth and twentieth centuries, vitalistic hypotheses of senescence characterized by the loss of a specific "vital force" were abandoned and gave way to attempts at renewing older Aristotelian-Galenic concepts under a modern form. German hygienist Max Rubner (1854–1932), for instance, carried out a series of very interesting experiments on the relation between ageing and metabolism and came to the conclusion that ageing was, in fact, due to a slow diminishing of metabolic processes in the protoplasm and a gradual extinguishing of vital energy.[16] The character of metabolism changes as well, since the proportion between anabolic and catabolic processes gradually worsens. The fall of basal metabolism with advancing age was established by A. Magnus-Levy (1865–1955) through observations on his own body, while experiments made by Rubner, Du Bois, Benedict, Boothby, and others have shown beyond doubt the correlation between ageing and the diminution of metabolism—i.e., the lowering of heat production through the diminution of oxidizing processes.

However, on the basis of these experiments, it cannot be stated whether ageing is the cause or the result of the diminution of metabolism. It still remains to be determined whether the diminishing of metabolism produces ageing or whether some primary ageing processes produce the diminishing of metabolism. Unfortunately, by experiment, only the correlation but no causal relation between these two phenomena can be definitely established. The same difficulty had to be faced when dealing with other hypotheses concerning ageing.

Vitalist hypotheses relating to the wasting of "the innate stock of vital energy" inspired analogous materialist attempts to explain the etiology of senility. Thus, Otto Bütschli (1848–1920), writing in 1882, conceived of ageing as the result from the consumption of a certain "vital ferment," admitting at the same time that absolutely nothing was known about the chemical composition and properties of this ferment.[17] A further modification of this idea was proposed by the American physiologist Jacques Loeb (1859–1924), who claimed that ageing is a gradual loss of a certain chemical substance, given in certain measure to each individual at conception.[18] This substance, Loeb said, is consumed by metabolism and its loss cannot be avoided. In support of this hypothesis, he assumed further that life span is an inherited property and, as a consequence, that it must be represented in the germ cells. This argument does not carry much weight, however, given that what is inherited is not the amount of a given chemical substance supposed to maintain life but a complex biological structure. The inherited constitution may, to a considerable degree, influence the length of life, but this does not entail that longevity depends on the amount of a certain substance contained in the organism. Loeb's view is too simplified, but it should be admitted that his essay "Ueber die Ursache des natürlichen Todes" ["Natural Death and the Duration of Life"] published in 1908 had already anticipated some later discoveries concerning the function of hormones.

The supporters of mechanist-materialist theories of life, advocating the idea that an organism is like a special physico-chemical machine came, by analogy, to the conclusion that ageing was a wearing-out process (*Abnutzungsvorgang*). Like a mechanical device gradually worn out with use and the passage of time, an organism gradually grows more and more inadequate to carry out its functions. As a matter of fact, this concept goes back to very old times; it corresponds to some general ideas put forward by ancient philosopher-anatomists (Democritus of Abdera, Epicurus, Lucretius, and others) and physicians of the Methodist School (Asclepiades of Bithynia, Themison of Laodicea, Soranus of Ephesus, and others), although in the preserved fragments of their works no direct account of ageing can be found.

In the seventeenth century, the mechanistic concept of ageing was supported by the followers of the Iatrophysics School, such as Giovanni Alfonso Borelli (1608–1679) and Giorgio Baglivi (1668–1707). This concept was at its height at the close of the nineteenth century, as it was advocated by biologist Richard Hertwig (1850–1937), physiologist Max Verworn (1863–1921), clinician Bernhard Naunyn (1839–1925), and others. But all these authors are rather vague about how an organism wears out. For the supporters of this hypothesis, a general analogy between ageing of the organism and the wearing out of a machine continuously in use seems to suffice.

A number of clinical and patho-anatomical observations, however, run counter to the hypothesis of ageing as a process of wearing out with use. The behavior of an organism is opposed to that of a machine: Activity regularly produces growth of both the living system and its functional capacities, while inactivity leads to atrophy. This fact has long been known about muscles, but Oskar Vogt (1870–1959) has recently established it with regard to the vitality of ganglion cells. As a matter of fact, on the basis of his vast experience in autopsy, Julius Friedrich Cohnheim (1839–1884) has given evidence that the degree of senile involution is very little influenced by the physical strain exerted upon the organism in the course of its life.[19]

Senescence as Surplus: Experimental Evidence

Senescence need not be regarded only as the result of a loss; the surplus of a noxious substance, the intoxication of the organism can also account for it. Hints at this possibility can already be found in ancient and medieval physicians. Renaissance revolutionary in medicine Theophrastus Paracelsus (1493–1541), for example, looked upon intoxication as the cause of senescence. An original hypothesis was also suggested by Russian biologist Élie Metchnikoff (1845–1916), Pasteur's collaborator, renowned for his theory of phagocytosis. Metchnikoff defended his attitude very forcibly and his *Optimistic Studies* were a sensation in the first decades of the twentieth century.[20] His work attracted much attention from physicians and still more from laymen. According to him, senescence was due to a slow, chronic intoxication of the organism by some specific toxins produced by intestinal bacteria as their metabolic products. Bacterial toxins, on the one hand, have a deleterious effect on the vital functions of delicate tissues (on nervous cells in particular), while on the other hand, they intensify the activity of phagocytes that in turn damage and ingest other weakened cells. Senescence is the result of a pathologic phagocytosis process brought about by intoxication through the alimentary tract, which leads to a progressive atrophy of the central nervous system, the atrophy of other organs, to arteriosclerosis, and thus to senility and death. Metchnikoff did not think senescence was necessarily a physiological process. In his opinion it was an accidental disease, which could be controlled by preventing intestinal putrefaction. To this end, he recommended yogurt, *Bacillus bulgaricus*, and related bacteria having an antibiotic effect on agents of putrefaction. Metchnikoff even suggested a radical surgical extirpation of the large intestine as a preventive measure against "the disease of senescence"! In light of modern scientific investigations, Metchnikoff's assumptions turned out to be wrong. In senescence neither exogeneous intoxication nor pathologic phagocytosis play any part whatsoever.

A further hypothesis of intracellular sedimentation or internal intoxication was formulated, according to which senescence is the consequence of the accumulation of harmful waste products of the metabolism of the cells (M. Mühlmann, R. Pearl, R. Rössle, C. & O. Vogt, and others). These ideas were greatly strengthened by Alexis Carrel's (1873–1944) experiments, started in 1911. With in vitro cultures of tissues in which waste products were constantly washed away and fresh, nutritive substances added, he succeeded in lengthening quite considerably the life of certain tissues of the chicken embryo, in comparison with the life span of a hen. If, in a culture of tissue waste products are not washed away, Carrel noted, there soon occurs a retardation of the growth and the self-intoxication of the cells. From this observation, he drew a farreaching conclusion—namely, that all cells are potentially immortal and that senescence and death are due to changes in the chemical composition of body fluids brought about by insufficient washing away of metabolic by-products. He provided evidence that the serum of the aged prevented the growth of cultures of tissues, whereas the blood of the young enhanced the growth.[21] Carrel's ideas of life, ageing, and death are well-known to the general public by his book *Man: The Unknown*, written in a very interesting and popular way.[22]

The hypothesis of self-intoxication has a fairly solid experimental basis. However, it should not be forgotten that these experiments, once again, established only a correlation between certain phenomena but produced no definite explanation concerning their causal relation. We still do not know whether an increased amount of certain harmful substances in the senile serum is brought about by some primary process of senescence or whether the accumulation of these substances produces the processes of cellular senescence. Experiments have shown the existence of some internal poisons in aged organisms, but it remains to be explained whether they are the cause or the symptom of senescence.

Full merit should be given to experimental biology for clarifying a very important point—namely, that the diminution of the power of regeneration of an organism is one of the basic features of age changes. Istrian physician Santorio Santorio (1561–1636) called attention to this fact.[23] Among his aphorisms concerning "static medicine" (i.e., the quantitative analysis of metabolism) the following one is especially worth mentioning: "Why many children live longer than old persons? Because, he says, they may be often renewed, from the lowest standard of weight to the greatest."[24] In aged persons, metabolic processes proceed at a slower rate, and this is why perspiration decreases, bodily tissues become rigid, and the power of regeneration gradually dies out. Since Santorio also tried to explain the etiology of certain diseases by means of changing metabolism and perspiration, it is natural that he should have come to the following conclusion and to his subsequent therapeutic deductions: "Old age may truly be reckoned a distemper, but it may be long protracted if the body perspires well."[25] Santorio's contemporary, the unhappy English chancellor and outstanding philosopher Francis Bacon (1561–1626) also observed in his book *Historia vitae et mortis* as follows: "Repairing [of the body] proceeds from the unequal repairing of some parts sufficiently, others hardly and badly in age."[26] That is to say that Lord Bacon also thought that old age was the result of the diminution of the power of reparation.

More recently Alexis Carrel and Pierre Lecomte du Noüy (1883–1947) have studied the healing of experimental skin injuries in animals and humans of various age. They demonstrated that the rate of healing is the mathematical function of age. Thus, the epithelization of a sterile surface injury of 20 cm^3 in size in a ten-year-old child takes about twenty days, and about one hundred days (that is, five times longer) in a sexagenarian. There are, however, certain individual variations in the speed at which individual organisms age. It may be assumed that there is an internal biological time that is not quite identical with physical time. Lecomte du Noüy is of the opinion that the rate of healing of an experimental injury (that is, the rate of speed of regeneration processes) may be used as a criterion to ascertain the biological age of the organism.[27]

In the period between 1890 and 1908, American embryologist Charles Sedgwick Minot (1852–1914) formulated a fundamental biological hypothesis according to which ageing is due to the diminution of growth and the power of regeneration.[28] One of the most characteristic age processes, he argued, is the change in size ratio between cellular protoplasm and the nucleus (cytomorphosis). With advancing age, cellular protoplasm grows in size in relation to the size of the nucleus. Metabolism and cell regeneration are most intense in the period of embryonic growth, but these processes are continued later in life, although in a quantitatively different way. Senescence, thus, is qualitatively the same biological process as embryonic development and death results from a high differentiation of an organism's cells. Minot's ideas were supplemented by German pathologist Robert Rössle (1876–1956) in his very illuminating work on the relation between growth and ageing.[29]

Carrel's experiments with in vitro tissue and organ cultures also gave strong impetus to the idea that ageing was due to the common life of numerous cells and inevitable only for the organism as a whole, while individual cells were potentially immortal and could remain eternally young. August Weismann (1834–1914) was the first to provide a scientific basis for this assumption.[30] In his opinion, monocelled systems are potentially immortal and do not age but die of a violent death by some external accidents. Senility and death are characteristic of metazoa, multicellular organisms only, but even in them a component, *the germ plasm*, as Weismann calls it, is potentially immortal and does not undergo ageing processes. Weismann's division of the organism between the immortal germ plasm and the inferior soma, reported in 1882, aroused much discussion, especially among geneticists. At present, his division is neglected since, on the one hand, if no special process of fecundation occurs, the germ cell dies just as well as the soma and, on the other, it has been shown by the study of cultures of tissues that even a somatic cell may be declared potentially immortal. The hypothesis according to which senescence results from the common life of a number of cells may be extended by the further assumption that it results from a disharmony in the development of individual cells, tissues, and organs. Metchnikoff, too, mentioned the rivalry among cells as an etiological factor of ageing. The significance of the disproportion in the development was adequately pointed out by Rössle in his work mentioned above.

Senescence as Involution: Historical and Contemporary Approaches

Senescence may also be considered as the consequence of the primary involution of certain tissues or organs. This, in fact, is a practical formulation of the idea of the disproportion of individual components in multicellular organisms, which was scientifically supported by cytologic and histopathologic investigations in the second half of the nineteenth century, although similar ideas, naively formulated, can be traced back to ancient physicians who thought that the heart was the source of life and the center of "innate heat." Here, we can see the ancient medical antinomy between organismic and particularistic attitudes toward physiological processes—continuing from the time of the conflict between the Schools of Cnidus and Cos, to the iatro-physical and iatro-chemical theories, and up to Rokitansky and Virchow and most modern scientific controversies—transpiring once again. Against this ideological background, we can easily understand why it was Hippocrates and Galen who thought that an organism aged as an entity and the anatomists of the School of Alexandria who made a single organ responsible for it.

In 1597, André Du Laurens (1558–1609) wrote that Egyptians and Greek physicians in Alexandria believed that the natural cause of old age was the diminishing of the heart, which, up to fifty years, gains two drachms in weight every year and then gradually becomes lighter. Du Laurens disagreed with this view, as he weighed the hearts of the elderly and found them as heavy as those of the young.[31] The hypothesis of the primary failure of heart and blood circulation was revised in a modern form in the nineteenth century when a number of outstanding clinicians and some patho-anatomists made arteriosclerosis, particularly the arteriosclerosis of coronary and brain vessels, responsible for ageing. "The man is as old as his arteries" ("On a l'âge de ses artères"), as Henri Cazalis put it aphoristically.

Valli, in turn, looked upon senescence as the result of calcium accumulating in soft organs. Recent investigations by means of isotopes (^{45}Ca) have shown that the absorption of calcium in soft organs of aged animals is really accelerated, while its metabolism is lowered. The idea of arteriosclerosis as a primary senescent process, however, is now abandoned. Most contemporary patho-physiologists and patho-anatomists envisage it as a morbid complication, which does occur more frequently in the aged but cannot be regarded as the cause of "normal senescence" either.

From ancient times, the brain was considered one of the noblest organs. Alcmaeon of Croton (c. 510 BC), for example, believed that the brain was the center of sense perception and the seat of memory and thinking. From the ancient Greek world up to the nineteenth century, it was now and then pointed out that senescence was primarily the result of brain changes. But these observations were tentative; they were suggestions rather than proper hypotheses. At the end of the nineteenth century and the beginning of the twentieth century, however, a series of remarkable scientists (e.g., Mühlmann, Metchnikoff, Ribbert, Rössle, Cerletti, and Vogt) accepted and expounded this idea, building up their systems on the morphological observations of age changes in nerve cells and the cytophysiological theory according to which neurons are the most differentiated and the most susceptible of all cells, having almost completely lost the power of regeneration. M. Mühlmann, for instance, gave an extensive explanation of the concept, advocated by many modern gerontologists, according to which senescence and natural death are due to a pigment atrophy of nerve cells caused by the accumulation of lipofuscin ("ageing pigment"), a harmful metabolic product.[32] The problem, however, is that, lipofuscin might be no waste pigment at all, as Rudolph Altschul (1901–1963) pointed out, but a useful substance assisting the functioning of the cells. It may be that lipofuscin accumulates in aged cells because, owing to their weakened functioning, they are no longer able to consume it. But here again it cannot as yet be said with certainty which of these two processes is primary: the accumulation of lipofuscin or the ageing of the cells.

Lately, various structural changes have been found in aged cells of the cerebral cortex—such as the diminishing of Nissl bodies in the cytoplasm, increased basophilia of the nucleus, amitotic division, etc.—but none of these changes can for certain be considered a primary process of senescence.[33] Russian physiologist Ivan Petrovich Pavlov (1849–1936), with his far-reaching analyses of conditioned reflexes, offered new possibilities to look upon ageing as a result of the functional derangement of higher nervous activity. His concepts of ageing were theoretically and experimentally developed by Russian scientists L. A. Andreev, L. A. Orbeli, V. K. Fedorov, Zh. A. Medvedev, and others.[34] It is worth noting that during the fourth Congress of the International Association of Gerontology at Merano in 1957, the Dutchman J. Groen set forth a theory according to which senescence results from the failure of regulatory functions of the diencephalon.

Also, according to Robert Rössle, ageing and natural death of higher organisms are due to a too advanced differentiation of nerve cells.[35] The quickest ageing can be observed in those tissues that, owing to their differentiation, quickly lose the power of regeneration. Thus, neurons age most quickly and mesenchyme most slowly. Senescence in humans, he thought, is thus the result of gradual degeneration of the central nervous system caused by the loss of the power of regeneration of neurons. This hypothesis is also open to criticism: If it were correct, a higher rate of speed in ageing should be expected in

organisms in which the differentiation of cells is greater, being phylogeneti-
cally on a higher level. Yet, this is not the case in nature. On the contrary, man
lives longer than any of the mammals (except elephant and rhinoceros) and
even longer than the majority of other animals.

In recent times, Soviet patho-physiologist Alexander Alexandrovitch Bogomolets (1881–1946) put forth a hypothesis quite opposed to the concepts
set up by Rössle and most other modern scientists, as he tried to demonstrate
that the reticuloendothelial system, which is the most active and the most
vital mesenchymal tissue, is responsible for ageing.[36]

After a sensational lecture delivered in 1889 by distinguished physiologist
and neurologist Charles-Édouard Brown-Séquard (1817–1894) at the Société
de biologie in Paris on June 1, human senescence began to be interpreted as
a result of a primary involution of sexual glands. Aged seventy-two at that
time, Brown-Séquard declared that after experimenting on animals, he had
for some time injected himself subcutaneously with extract of fresh testicles
of dogs and rabbits and felt rejuvenated, both physically and mentally.[37] The
actual value and significance of Brown-Séquard's observations, as of some
later similar experiments carried out by Eugen Steinach (1861–1944), Serge
Abrahamovitch Voronoff (1866–1951), and others, will be discussed in the
chapter on rejuvenation.[38] It was these experiments mainly that led to the
hypothesis of the primary significance of sexual glands and their products in
the etiology of ageing. The example of the estrus cycle of the female, covering
three distinctly different life periods, was also put forward in support of this
hypothesis. Senile involution of the female organism begins with menopause
and is evidently related to the endocrine hypoactivity of ovaries.

Twentieth-Century Perspectives on Involution and Senescence

In the twentieth century, speculation began about the primary involution
of some other endocrine glands being possibly responsible for senescence.
Thus, for example, Arnold Lorand singled out the thyroid gland, which in
the old usually atrophies and is impoverished in colloids.[39] Since H. M. Evans
and J. A. Long discovered the growth hormone in 1921, and P. E. Smith,
B. Zondek, and S. Aschheim the gonadotropic hormones in 1927, the central
role of hypophysis as a regulator was stressed, and this gland was assumed to
be "the biological clock" regulating the ageing of the whole organism. Roma-
nian gerontologist C. I. Parhon (1874–1969), for example, is of the opinion

that senescence is a dystrophic process, a derangement in nutrition and tissue correlation, which is mainly a consequence of hormonal perturbations. Parhon insisted on the role of the complex "hypophysis—adrenal gland—thyroid" and the influence of vitamins and similar substances on senescence.[40]

Experimental and clinical endocrinology have rapidly moved forward in the past decades, throwing more light on the genesis of some symptoms of senescence.[41] At the same time, however, it has become clear that fundamental gerontologic problems cannot be explained by hormonal factors alone. Although a number of old age symptoms have turned out to be correlated with hypofunction of gonads, thyroid and adrenal glands, and hypophysis in particular, all these are mere secondary phenomena. The atrophy of one or more endocrine glands cannot produce a state identical with senility. It has long been known that eunuchoidism and not senility is the consequence of castration. Hypophyseal cachexia (Simmonds' disease) is a state most similar to very advanced old age, but in spite of that they differ essentially. In sum, it is most likely that the involution of endocrine glands is not the cause but the consequence of a general senile involution of the organism. Reduced hormonal activity is only a beneficial form of adaptation to the general diminishing of cellular metabolic processes. Contrary to the enthusiastic belief of earlier experimenters, it has later been demonstrated that the stimulation of endocrine organs may be disastrous to the aged.

Hypotheses concerning the primary involution of certain organs are scientifically grounded insofar as certain organs can be considered *locus minoris resistentiae* to the destructive action of time, to the primary process of senescence. Such reasoning is useful to the clinic and the understanding of some specific patho-physiological processes going on in aged organisms, but only to the extent of shifting, for its supporters, the whole problem from the original question (What is the cause of the senescence of an organism?) to a further question just as difficult as the former (What is the cause of the senescence of the organ producing senility and death?).

Physico-chemical hypotheses go deeper into the essence of these problems. Many nonliving systems (e.g., colloid solutions, radioactive matters) are subject to "ageing," and it may even be assumed that the universe tends to an equilibrium of energy potentials, growing "older" and "deader" every moment. The idea of trying to find an explanation of the ageing of living organisms in some molecular processes—that is, in the identification of organic and physical ageing—may be heuristically justified by the fact that the living protoplasm is itself a physico-chemical colloid system.

It is interesting to note that when Max Rubner, in 1908, considered the possibility that ageing is the consequence of progressive dehydration of tissue colloids,[42] he resuscitated the ancient Galenic concept in a modernized form. It has already been said that Rubner was the one who provided evidence for the correctness of the ancient concept that ageing was a gradual waning of metabolic processes, of the inner body heat. While Galen thought that "innate heat" was slowly consumed through lack of moisture in the body, Rubner said that the diminishing of metabolic processes is the consequence of the dehydration of colloids. The analogy between these two hypotheses is striking. Yet, nobody, not even Rubner himself, nor later gerontologists, nor the historians of medicine, have taken notice of it.

Further formulations of the colloid theory of ageing have been given by Marinescu, Lumière, Ružička, and others. Thus in 1919, Gheorghe Marinescu (1863–1938) studied how protein colloids have the property of spontaneously and steadily diminishing their dispersion, so they become dehydrated. In his view, this is the cause of ageing in living protoplasm.[43] Auguste Lumière (1862–1954) offered evidence concerning the correlation between the coming to maturity of colloids and the ageing of the protoplasm.[44] By the flocculation of colloids he also tried to explain various pathologic changes, for instance anaphylaxis, immunity changes, inflammation processes, etc. As to Vladislav Růžička (1870–1934), he proposed the term *hysteresis* for the property of colloid solutions to change spontaneously into a more stable, concentrated state by which they lose the power of retaining water.[45] To explain ageing processes, much attention has also been paid to the physical properties of cell membranes. In recent times, T. B. Robertson, D. Reichinstein, and others have made original contributions to the physico-chemical explanation of ageing in living molecular and cell systems.[46]

All physico-chemical hypotheses of ageing formulated so far, however, are subject to strong criticism. For instance, they cannot explain why the molecular process of ageing does not affect the fertilized germ cell. Also, the analogy between ageing of living protoplasm and a nonliving colloid system has no justification, as the living matter is continually building up. The protoplasm is a dynamic system compared to nonliving colloid solutions, which are static. In a living organism, each molecule maintains itself only for a certain time and then is replaced by another. Only this turnover of the living matter is constant. The conditions in nonliving colloid solutions are essentially different: These systems consist of constant molecules changing only their mutual relation. Nonliving structures "age" unlike living organisms; they do not destroy themselves, nor do they rebuild themselves out of new component parts. It appears, therefore, very likely that the ageing of living matter, although similar to some degree to the ageing of inanimate structures, is regulated by some specific laws. Hysteresis of the colloids of protoplasm differs from hysteresis of nonliving colloid solutions, as the living matter is no closed system tending toward greater inner stabilization but an open dynamic system in which both hysteresis and other changes are in fact the consequences of an incomplete regeneration at the molecular level. The most recent investigations by means of marked atoms, radioactive tracers, have shown that the vital turnover is quicker than has hitherto been thought and that its rate decreases in aged individuals. Ageing can, therefore, be considered a specific biological process that is the result of insufficient and ever slower restitution of macroscopic and microscopic as well as molecular structures.

Ageing in Living and Nonliving Systems: The Contributions of Soviet and Eastern European Scientists

Soviet scientist A. V. Nagornyi and his collaborators (Bulankin, Nikitin, and others) have developed a complex chemo-biological theory of senescence. In the opinion of this group of scientists, senescence is due to the intracellular differentiation of living matter. Unstable protoplasmatic proteins do not regenerate completely but gradually develop more stable forms with an ever lower vital power. Ageing increases the proportion between metaplasmatic and active protoplasmatic structures. Nagornyi's theory has recently been criticized by Zhores A. Medvedev, who argued that ageing seems to be the result of some specific biological process of the inactivation of metabolism rather than the consequence of spontaneous physico-chemical changes.[47]

Slovenian patho-physiologist Andrej Župančič (1916–2007) supports similar ideas by saying that the structures of living organisms, in the course of their functioning, continually change, destroy, and reconstitute themselves at the same time and that ageing is, in fact, a gradual loss of the power to rebuild constantly changing structures. As time passes, the aged organism is less and less like itself, gradually losing the characteristics of a living system and, in its reactions, gradually approximating a state similar to that of nonliving systems.[48]

The twentieth century brought the revival of old concepts according to which ageing is due to a cosmic noxa; now, however, neither deities, in order to punish mortals, nor the sympathetic astrological influence of celestial bodies (a medieval concept: old age is the imprinting of the planets on the body [*senectus est planetarum impressio in corpus*]) have been evoked to account for it. Kunze, for example, proposes that ageing is the harmful result of cosmic and other ultraradiations that steadily shell living systems and destroy the nuclei of their cells.[49] A similar concept was set forth in 1957 by radiologist G. Failla and according to whom senescence may be due to the accumulation of irreversible lesions produced by spontaneous somatic mutations caused by different mutation agents, and particularly by ionizing radiation.

Senescence can generally be regarded as the consequence of the struggle between some harmful factors and the resistance of the organism. German physiologist A. Pütter gave a very interesting mathematical formula of this concept in 1921, which was further analyzed in two extremely instructive papers by biochemist K. Miescher and physician G. Schlomka.[50] Ageing and death in organisms, they argue, just as the destruction of radioactive matters, obey a similar mathematical law.

Most recent hypotheses connect ageing with physical laws on the irreversibility of some processes, with a physical tendency toward the increase of disorder among molecules. Thus, in 1924, Czech physiologist Vladislav Růžička (1870–1934) emphasized that hysteresis of protoplasm could be regarded as a consequence of the thermodynamic law on entropy.[51] Swedish biochemist Hans von Euler-Chelpin (1873–1964), in turn, recently reported that ageing was the consequence of the increase of entropy in hormonal reactions, especially in the macromolecules of hypophyseal hormones. *Entropy* is meant here as a term indicating anarchy in some processes, a tendency toward disorder. In his highly instructive book on the biology of senescence published in 1956, Alexis Comfort (1920–2000) says that the most plausible general theory of ageing is the one suggested by Bidder, according to whom ageing is no part of the inherent life pattern of an animal but is rather the anarchic situation resulting after that pattern has run its course. As Comfort expresses it: "Senescence is typically an indirect process, not a part of the programme, but a weakening of the directive force of the programme."[52]

W. Kuhn also connects ageing with changes in the synthesis of optically active matters of the organism. Owing to physical laws, in the organism the optical purity of certain matters changes, a progressive racemization takes place, while the presence of optical antipodes of certain matters essentially disrupts metabolism, hereby producing a harmful effect on the organism.[53] Many different hypotheses concerning the causes and the essence of ageing may be correlated and mutually support each other. A number of contemporary authors (e.g., S. Hirsche, A. J. Lansing, D. Kotsovsky, F. Henschen, L. Binet, M. Bürger, and C. I. Parhon), however, think that such a complex process as ageing cannot be ascribed to a single cause, such as the atrophy of sexual or other endocrine glands, the changes in the cells of the central nervous system, the intoxication by intestinal products, the accumulation of waste products of cellular metabolism, cytomorphosis and the decrease of metabolism and regenerative power, hysteresis of the colloids of the protoplasm, calcium deposition and arteriosclerosis of blood vessels, cosmic rays, the increase of entropy, and the like.

Most of these hypotheses contain at least a grain of truth, but none can solve the problem of ageing in general. Instead of attributing biological processes to a single cause, the idea of attributing them to many causes and conditions is gradually gaining ground, tending to embrace and synthesize a multitude of external and internal factors of ageing.

A Survey of the Mechanical Interpretations
of Life from the Greek Atomists to the Followers of Descartes

The first consistent mechanical theory of nature was formulated by the Greek atomists Leucippus (early fifth century BC) and Democritus (c. 460–370 BC) and developed by their followers, Epicurus (341–270 BC) and Lucretius (c. 99–55 BC). In their opinion, human and animal bodies are composed of atoms, and all life manifestations are only the results of complicated, but strictly determined, motions of these elementary particles. The heavy atoms composing the body, Greek atomists said, are moved by the soul. Yet this Epicurean "soul" is not a spiritual principle but a collection of very light and small atoms. Mind itself is an ensemble of extremely small atoms. Sensations are born out of movements and the soul provokes all bodily motions in a strictly mechanical way.

A strictly materialistic and deterministic biological doctrine was thus elaborated before the end of the fourth century BC and contained all the basic principles of mechanical materialism, with one exception: the analogy between organism and machine. But this is quite understandable given that complicated, self-moved machines had not yet been constructed and were not even imagined by ancient scholars.

There are, in fact, mechanical analogies in the Hippocratic corpus—for example, a comparison of the digestive organs with communicating vessels and a mechanical explanation of birth labor. Also, Plato (427–347 BC) compared vertebral movement to that of the hinge on a door.[1] But these analogies have not been followed up by philosophical generalizations.

In his famous treatise *On the Motion of Animals*, Aristotle (384–322 BC) compared the action of muscles and bones of the forearm with that of the classical weapon, the catapult.[2] Aristotle's use of such a model is only partially mechanistic, however, because this analogy serves to demonstrate the structural condition of a form of movement, not the origin and initial conditions of the movement. In ancient machines, the source of energy is provided by human or animal muscular work; the machine-organism analogy, thus, did not address the most fundamental issue—the origin of activity. In classical machines the motion is strictly passive; in living beings it is obviously active, if not necessarily spontaneous.

Aristotle's attempts to develop a mechanical model of organic phenomena are, in fact, vitalistic—that is, involving a special, nonmaterial vital principle. Even the very curious sentence in his *Politics* where a slave is called an "animated machine" is vitalist.[3] In Aristotle's mind, all movement requires a prime motor (mover), which must be immobile. It can thus be neither a muscle nor any other part of the animal body. The animal body is truly moved by desire, a strictly vital property that can only be explained by the soul. Aristotelian philosophy makes a fundamental distinction between animated and nonanimated motion. Incidentally, in Aristotelian physics the term *motion* has a broader meaning than in modern terminology. Here, motion is the translation of an object from one place to another and embraces other kinds of changes—such as creation and destruction, alteration, and increase and decrease. Nature, for Aristotle, cannot be explained only in terms of material translation. Every motion has not only a direction but also a purpose. The motion of celestial bodies is perfect and eternal, whereas terrestrial motions are only a search for perfection.

Erasistratus, a great physiologist of the Alexandrian School (third century BC), has proposed a coherent mechanical theory of life. In his view, the human body is held together and moved by three systems of networks: arteries, veins, and nerves (which were thought to be hollow). According to this system, arteries contained "pneuma"; veins contained blood; and nerves contained a special kind of fluid called the nervous spirit. All vital functions of the body are elicited and controlled by these three systems of organismic integration.

For Erasistratus, an animal or a human body is a hydraulic and pneumatic machine, although such a machine was not yet known. It is probably not a mere coincidence that the first hydraulic and pneumatic machines were constructed by two Alexandrian scholars, Ctesibius (end of the third or beginning of the second century BC) and Heron (about the beginning of the first century BC). The work of the latter is also an interesting testimony of Greek theaters of automata.

Alexandrian technologists were responsible for inventing a force pump, a water clock, and a special organ, a musical instrument in which the air, instead of being supplied by human lungs, came from a wind machine. Action of heat on water was used as motion power in other machines. This was particularly important as it could explain how innate animal heat could produce movement. At the time when Rome was at its height, the mechanistic theory of life was accepted by a particular medical sect called the Methodists, founded by Asclepiades (c. 150–65 BC) and Themison (first century BC). Methodists followed the teachings of the atomists in considering the human body as a network of fibers and pores between which atoms of various shapes circulated.

The anatomical work of the Alexandrian School, together with the Aristotelian notion of the organs, have survived in Galen's anatomo-physiology, and especially in *On the Usefulness of the Parts of the Body*,[4] a treatise in which he offers mechanical explanations of the structure of some organs, such as the articulations of joints. His description of the hand as a perfect mechanical tool is a masterpiece. Galen's explanations, however, are not materialistic. On the contrary, the main purpose of his anatomical and physiological writings was to establish the existence of a directive spiritual force, which was a final and not a causal determination.

Biomechanical Analogies at the End of the Middle Ages

During the Middle Ages, nothing particularly important for our subject happened. It was only at the end of this period (i.e., in the thirteenth to the fifteenth centuries) that new kinds of machines were constructed. By then, it was understood that a machine could command others, but this phenomenon of a "life of relation" had yet to find an echo in biological thought. The same lack of interest followed Villard de Honnecourt's construction of a mechanical "eagle." The wonderful achievements of Leonardo da Vinci (1452–1519) should not be overlooked, but his biomechanism was Aristotelian (i.e., nothing more than a relatively simple mechanical imitation of some movements of articulations and extremities). Leonardo's study of the mechanics of phonation, however, was curious and far ahead of his time. Biomechanical analogies continued to appear sporadically in the new anatomy of the sixteenth century and their influence can be perceived in the title of Vesalius's book *On the Fabric of the Human Body*.[5] With Rabelais (1494–1553), also, anatomy is filled with mechanical metaphors: filters, press, cordage, pulleys, etc.

A Spanish physician, Gómez Pereira (1500–1567), attempted to demonstrate that animals lack sensitivity and are thus very different from humans, who have sensitive, intelligent, and immortal souls. Pereira's views were introduced in a rare and curious book entitled *Antoniana Margarita; opus nempe physicis, medicis, ac theologis non minus utile quam necessarium* published in Medina del Campo in 1554. Sensing how much his approach was new and paradoxical, he chose a meaningless title for his book. Actually, he gave it the name of his father and his mother, Antonio and Margarita! This was most likely a kind of camouflage intended to protect himself from the Inquisition. For the same reason, he dedicated his work to the Archbishop of Toledo.

Pereira denied that animals have any kind of spontaneity. His basic statement was that "animals do not feel" (*bruta non sensire*).[6] Of course, he did not disagree that a dog or a horse can "see" or "smell" his master or his enemy, which makes it move toward his master or away from his enemy; but he argued that these movements occur automatically and not, as in man, through the intermediary steps of psychic experience and mental propositions.

Pereira distinguished four modes of animal movements: First, present objects (*rebus praesentibus*)[7] can act directly on their sense organs and elicit immediate responses from animals. Second, animals can move because the contents of their memory, the "image of" things (*phantasmatibus rerum*),[8] can also prompt movements. It is important to emphasize that for Pereira these "images" are real, corporeal emanations of things and are not illusory. Such *phantasmata* are indeed the very cause of animals' search for food and other similar movements. In fact, the first and the second of Pereira's modes somehow mirror Ivan Pavlov's distinction between unconditioned and conditioned reflexes.[9] Third, animals can move by others' command or "from the aforesaid" (*ab altero praedictorum*),[10] as in the training of dogs or parakeets. And fourth, they can move by natural instincts (*ex instinctu naturae*)[11] as, for example, the behavior of bees and ants. Here is the main issue because, using the example of digestion, Pereira included in this group all internal movements of animals that we call physiological.

Even if Pereira's basic assumption that there is a fundamental difference between animals and man, only the latter having a soul, the Spanish physician still maintained that there is no basic distinction between physiological processes in both kinds of living beings.[12] In both, he said, internal movements are automatic and directed by some kind of instinct. Pereira was very cautious, however, and the machine-organism analogy cannot be found explicitly in his work. There are no references to mechanical constructions but, without doubt, his book contains at least implicitly the idea of organic disposition as the only basis of animal external behavior and animal and human physiological processes.

But Pereira's concept of life is not mechanistic in the modern sense. His philosophy is still governed by the notion of "hidden qualities" [*qualitates occultae*]. Thus, we can understand why the only explicit analogy found in his work is between animal behavior and magnetic attraction. Somebody without historical perspective could think he anticipated the tropism theory.[13]

Pereira's paradox evoked no interest, however. He was attacked by a theologian from Salamanca, Miguel de Palacios, but he maintained his position and, in his response, used, it seems, even more clearly, the notion of beast-machine. Descartes was accused of having been inspired by Pereira's work but denied it.[14]

The first edition of Pereira's work was hardly available to Descartes, but the possibility of a direct or indirect acquaintance with the second edition of Pereira's book (1610) cannot be excluded. Descartes, just like Pereira, spoke of his hypothesis on animal automatism in terms of a paradox. Anyway, we shall see that Descartes pressed the machine-organism analogy much further and that the question of priority is of no significance in this case.

Descartes and the Revival of Mechanical Biology

When Descartes expressed his views one century later, the philosophical background of European scientists was quite different and his explanations of living organisms found an astonishingly receptive audience. Understanding the revival of mechanical biology requires looking at the foundation of the new science of mechanics. No wonder that it was precisely the seventeenth century that witnessed the first mechanical theories of life in the modern sense. Under Galileo's influence, a special school of medical thought was born in Italy, the so-called Iatrophysics or Iatromathematics. The new laws of mechanics, the introduction of quantitative experimentation, and the use of the microscope gave to the followers of the Iatrophysics School the impression that all physiological phenomena could and must be explained in terms of simple physical properties of vital structures and movements. Two great achievements of this Galilean orientation in physiology were Harvey's discovery of blood circulation (1628)[15] and Santorio's fundamental experiments on metabolism (1614).[16]

Nowadays, chemical and physical interpretations of life are considered to be complementary with one another, not contradictory. But this was not the opinion of ancient scholars for whom physical and chemical interpretations of life stood in opposition. Mechanistic interpretations tried to explain all phenomena in terms of the motion of body particles that were subjected to strict mechanical laws, neglecting or even denying the essential properties of matter. In contrast, chemical explanations of life considered "hidden qualities" of matter to be of primary importance. Today, mechanical models of life are often regarded as the simplest ones. Surprisingly enough, this apparent simplicity does not indicate a chronological or historical priority. Mankind began working with very complicated, animistic and later pneumatic and humoral explanations, which certainly were more chemical than physical. Then, from classical antiquity until the Enlightenment a progressive simplification of biological doctrines—that reaches its peak with the faith in a complete reduction of animal and human organisms to very rough mechanical devices— is noticeable.

Seventeenth-century European scholars acquired knowledge about how to construct new types of machines in which parts responsible of the regularity of movements were dissociated from parts generating energy, thus reinforcing the illusion of a kind of autonomous life. Hydraulic automata built in the royal gardens at Saint-Germain-en-Laye, near Paris, are good examples of such automatic machines. A well-established historical tradition recounts that it was precisely the witnessing of this mechanical microcosm in the king's gardens that gave to the young French philosopher René Descartes, in a sudden illumination, the idea of the beast-machine.

The first printed account of Descartes's fundamental hypothesis is found in his very famous *Discourse on the Method* (1637). Here, Descartes said:

> This will not seem at all strange to those who know how many kinds of automatons, or moving machines, the skill of man can construct with the use of very few parts, in comparison with the great multitude of bones, muscles, nerves, arteries, veins and all the other parts that are in the body of any animal. For they will regard this body as a machine which, having been made by the hand of God, is incomparably better ordered than any machine that can be devised by man, and contains itself movements more wonderful than those in any such machine.[17]

In his *Treatise on Man*[18] (probably written about 1632 but published only thirty years later), Descartes asks the reader to imagine a clay machine similar in all respects to the human body. Considering the heat located in the cardiac cavity to be a force of motion, he then goes on to explain the overall functions of this machine by the simple laws of mechanics before leaving him wondering whether there is any difference between an artificial machine and a human being, except for the power of reason.

At first glance Descartes's position is a strange one, combining a strict mechanistic approach with a dualistic, idealistic philosophy. In Descartes's philosophy, the world is composed of two substances—the passive matter (*res extensa*) and the thinking soul (*res cogitans*). The soul is the essential part of the human being and Descartes was even convinced that he knew its exact location. Descartes believed that the soul is located in the pineal gland because this is the only important part of the brain that does not come in pair and that is not divided into right and left parts. The principal function of the soul is to think; as we are obviously not "thinking" our physiological processes, the soul cannot act at the physiological level.

Cartesian philosophy regards animals as having no soul, an absence their vital reactions do not suffer from. In a letter to Henry More (1614–1687) dated February 5, 1649, Descartes insists on a comparison between animals and mechanical automata: "Since art copies nature, and people can make various automata which move without thought, it seems reasonable that nature should even produce its own automaton, which are much more perfect than artificial ones—namely the animals."[19] The most important difference between a nonliving automaton and an organism, he writes, is the very special character of living beings' central power. As Descartes explains further: "I do not deny life to animals, since I regard it as consisting simply in the heat of the heart."[20] Thus, heat or fire is the first mover of living machines; their primary source of movement is the heart, or more precisely a kind of chemical process going on in cardiac cavities. Gilson[21] demonstrated very precisely the extent to which Descartes remained under the influence of scholasticism—which he so strongly opposed at once in his epistemology and mechanics—in accepting the classical notion of "innate heat."

If a human or animal organism is viewed in a Cartesian way, that is, as an automaton, the logical necessity to assume a divine intervention by the First Engineer cannot be avoided: A complicated machine must be built by some higher intelligence.[22] Two possibilities can then be envisaged. Let me express these in modern language: An animal-machine is either an automaton made of cybernetic regulations, something like a program-tape inserted in it by the First Engineer; or it is a kind of car or, better, a very complex factory, which cannot operate without permanent intelligent supervision. Descartes chose the first logical possibility, which was certainly very audacious on his part, for nothing was known of feedback circuits and computer programs in his time. The reason why he was unable to clearly unravel the implications of his beast-machine analogy is that he was looking for a still nonexisting mechanical model.

Let me emphasize some cyclical features in the history of mechanical concepts of life: Like Erasistratus many centuries before, Descartes was intellectually advocating the concept of a device that could only be realized by engineers of the future. At certain critical stages in the development of so-called exact sciences, I believe biology can offer foresight into the next step because biologists are both challenged and inspired by the contemplation of the highest, real, existing structural systems. The beginnings of cybernetics and the modern philosophical relationship between life and electronics is another important example. And this is certainly not the end of the story.

Promises and Challenges of Mechanical Perspectives in Living Systems

Descartes compared the neuromuscular system of his anthropoid machine with a church organ and its sets of pipes. This analogy was imperfect, but he was unable to find a better comparison for organisms taken as wholes than the working of a clock. As Descartes was fully aware of the limitations of his analogy, he pointed out that the dexterity of animals "proves rather that they have no intelligence at all, and that it is nature which acts in them according to the disposition of their organs. In the same way a clock, consisting only of wheels and springs, can count the hours and measure time more correctly than we can with all our wisdom."[23] A careful reading of this statement shows that for Descartes a clock can only partially be compared with a living machine.

Actually, one of the shared characteristics of the Cartesian beast-machine concept and a clock is the existence of a central and single source of energy. In one of my recent publications I tried to show how this assumption acted against the development of the fiber theory of organism in a direction that would later be favorable to the foundation of the cell theory.[24] Fortunately, this negative influence of Cartesian views was counteracted by the development of the biological concept of irritability, which was created by Glisson but more generally accepted only one century later (i.e., in Haller's time). Experimental evidence of the *vita propria* of animal parts (for example, the contractions of a freshly excised muscle as demonstrated by Stensen, Willis, Swammerdam, Baglivi, and others) was the principal reason why, for professional anatomo-physiologists, the Cartesian model remained an attractive but unacceptable hypothesis. Paradoxically, the most important immediate followers of Descartes's physiological doctrine were pure philosophers and mathematicians like Mersenne (1588–1648) or Malebranche (1638–1715).

Sixteenth-century life sciences were dominated by descriptive methods of research and especially by the description and classification of species and anatomical investigations. Conrad Gesner (1516–1565) and Andreas Vesalius (1514–1564) are prominent representatives of this type of biological research. In the seventeenth century, however, physicians and biologists wanted to explain the processes of life and not just describe their material basis. The most important biological discovery of this period, Harvey's theory of blood circulation, was certainly more than an anatomical description: It was a new interpretation of functions, of internal dynamics. But Harvey's theory was also profoundly connected with anatomical thinking; it was a kind of "animated anatomy"[25] (i.e., "anatomy in motion," as Haller called it).[26] Harvey's discovery, in fact, resulted from the fortunate combination of two new methods, both linked with the school of Padua where Harvey studied medicine. The first one, the Vesalian direct observation by a specific kind of dissection, was known to Harvey through the teaching of his master, Fabrizio d'Acquapendente (1537–1619). The second one was the Galilean quantitative method, a new conception of mechanics. Harvey gained knowledge of venous valves using the first method, but it was only in applying the second one that a correct interpretation of the circulation of the blood became possible. Harvey's discovery was not only that blood moved in closed circles but that its motor force was a pump.

In Harvey's physiology, the origin of movement was muscular contraction, a mechanical force in the strict sense. Descartes, one of the first great authorities to accept Harvey's theory of blood circulation, refused to consider the heart simply as a pump, however. In Descartes's account, the heart was a heat engine, something like an explosion motor moved by periodical expansions of blood in ventricles and elicited by innate heat. For geometrical reasons (dilatation of heart when beating against the chest wall), and accepting the wrong medieval view of the synchronicity of the radial pulse and the diastole, Descartes was convinced that the expulsion of blood coincided with the diastole.

Certainly, if the heart is a pump, and it is one, its active phase is the systole. Harvey easily demonstrated that Descartes was on the wrong path.

The major idea of Descartes's physiology (i.e., the analogy between an organism and a complex mechanical device) was quite enticing, and its success can easily be understood. But it must be emphasized that in terms of concrete details, Descartes's theory was merely a series of shortcomings. Successfully applied in mathematics and physics, his deductive method was definitely misleading in biological sciences.

The human mind cannot deduce living phenomena from a few general principles. In this field of science, the Galilean and Baconian method of induction followed by experimental verification of hypotheses was the only suitable way to make progress. Thus, the Iatrophysics School in Italy and England followed this experimental path and not Cartesian speculative philosophy. In spite of their common search for a mechanical model for life, Cartesianism and Iatrophysics have very different methodological approaches and should not be mixed together, as textbooks on the history of biology commonly do.

Galileo himself constructed physical models to explain some biological aspects of fish physiology. His explanation of a fish's swimming bladder in hydropneumatic terms was a good example of such a biological mechanism. The most important scholar of the Iatrophysics School, however, was Giovanni Alfonso Borelli (1608–1679), a physician and mathematician of Naples. His book *On the Movement of Animals*,[27] published posthumously in 1680–1681 (probably written about 1660), was the first practical application of mathematics and mechanical principles to the study of muscular work. His mechanical interpretations of the walking of man and quadrupeds, the swimming of fish, and the flying of birds are all excellent. Here, he tried to explain by mechanical calculations the movement of internal organs, for example, the pressure of the stomach on food and the pump activity of the heart. He stated that the steady flow of blood from the arteries into the veins was the result of the elasticity of the arterial walls. Borelli failed, however, to obtain correct numerical results. The same happened to Johann Bernoulli (1667–1748), the mathematical genius who introduced infinitesimal calculus into physiology.

In spite of the scientific approach, which we now consider to be essentially correct, numerical results obtained by the followers of Iatrophysics in the seventeenth century were quite unsatisfactory. The basic method was correct, but many observational data were too roughly determined. It was only with the development of physical measurement instruments in the second half of the nineteenth century that the dream of Iatrophysicians of mathematically analyzing muscular movement was realized.

Borelli's method differed from Cartesianism in three major ways: (1) his approach was empirical rather than speculative; (2) he rejected the theory of innate heat; and (3) he postulated the intervention of the soul in physiological processes. By a very simple and effective experiment, Borelli demonstrated the fallacy in Descartes's conception of the cardiac localization of animal heat. Measuring the temperature of the heart and other internal organs of a vivisected deer with a thermometer, he discovered that the cardiac cavity was no warmer in those places than in the other organs of the animal. Thus, it was proved beyond all doubt that the heart is a muscular pump—a mechanical device—not a thermic machine.

In spite of his mechanistic views, Borelli refused to consider living organisms as fully determined automata. For him, the animal body was not a clock, whose working could be strictly regulated mechanically. The Borellian model of an organism was that of an animated machine whose activity was the result of the permanent intervention of the soul. To express this idea in anachronistic terms of our modern technology: His theoretical model was not a robot but a complicated car with a spiritual driver; this driver, however, was not identical with our intellectual principle but with a vegetative soul.

In Cartesian philosophy, the creator of organisms was supposed to explain the teleological arrangement of their parts, but after the first act of creation the intervention of an intelligent principle was considered superfluous. In Cartesian physiology, God and soul were postulated then immediately forgotten. In the following centuries, this attitude led to an uncompromising mechanical materialism, which denied even divine creation. As in the time of Greek materialism, the apparently meaningful and purposeful behavior of organisms and the harmony of their structures were explained as being the result of natural selection. But this was really possible only after Darwin's patient and marvelous work.[28] Between the time of Descartes and that of Darwin, wholly mechanistic attempts to explain living phenomena were carried out to naïve extremes. Such explanations plainly oversimplified reality and the conclusions often expressed nothing more than a faith in natural theology, as expressed in the title of John Ray's book, writing at the end of the seventeenth century, *The Wisdom of God Manifested in the Works of the Creation.*[29] We will see that in the eighteenth century, the machine-organism analogy served not only as the basis for materialistic thinking but, at the same time, for most vitalistic theories like Stahl's animism. If the world-famous Danish bishop and anatomist Niels Stensen (1638–1686) described the brain as a "marvelous machine"—of which it is essential to know the anatomical disposition—his interest in it was also justified by his belief that the brain "is the main organ of soul (1669)."[30]

The general ideas of the Iatrophysics School were exposed in a most brilliant and provocative way by Giorgio Baglivi (1668–1707), a physician from Dubrovnik in Dalmatia. In his book, Baglivi compared the jaws to pliers; the stomach to a retort; the veins, arteries, and blood vessels to hydraulic tubes; the heart to a spring; the viscera to screens and filters; the lungs to bellows; and muscles and bones to a system of cords and pulleys (1696).[31] As Georges Canguilhem (1904–1995) pointed out, from the physical point of view these comparisons are not of the same order.[32] Cords and hydraulic tubes are mechanisms of transmission, whereas pulleys are mechanisms of transformation of movement; only the spring is a motor—that is, a device capable of originating movement. The central role of the heart is obvious in all mechanical explanations of life; Baglivi and Pacchioni tried to explain the role of the brain in a similar mechanical way, which, they assumed, worked by the contraction of its membranous envelopes. This obsolete theory needs no further comments, but we should insist on a particular aspect of Baglivi's thought: his doctrine of the living fiber. According to Baglivi, all physiological and pathological processes depend on fibers, which are some kind of elementary machines.[33]

*Chemical and Mechanical Interpretations of Life
in the Seventeenth and Eighteenth Centuries*

In eighteenth-century biology, a scientifically correct mechanical model of the living body had to consider the fact that the organism has in all its parts (or, at least, in most of them) local and to some extent autonomous sources of energy. Baglivi ventured a purely physical solution of this problem and failed. It was thereafter considered that no model of this kind could be conceived without also appealing to chemical phenomena. In one way or another, practically all biologists of the eighteenth century used the old theory of Willis, which explained the origins of movement in the living body by local chemical "explosions" in the muscles.

Chemical interpretations of life are older than mechanical ones. Ancient Hippocratic and Galenic humorism, for example, are in some way chemical theories of life. But in a narrower sense, the representation of living phenomena in technical terms of chemistry only began during the first half of the sixteenth century with the work of Paracelsus (1493–1541), who applied technical knowledge of alchemy to understand the functioning of organisms. For him, all living beings are composed of sulfur, mercury, and salt. These Paracelsian constituents of the animal body, however, were not chemical elements in the modern sense; rather, they were conceived of as three functional principles. Paracelsus and his followers—Jan Baptista van Helmont (1580–1644) and Franciscus Sylvius (1614–1672), supporters of Iatrochemistry—studied vital phenomena using chemical methods and assumed a perfect analogy between vital functions and chemical reactions taking place in retorts. Fermentation, thus, was considered to be a fundamental vital process. If Iatrophysicians brought about a more accurate scientific account of the knowledge of animal motion, circulation of blood, the mechanics of respiration, etc., the Iatrochemical School was able to better explain processes such as digestion and the origin of animal heat.

It is a curious fact that until the eighteenth century the chemical vision of life was always linked to intelligent spiritual principles, alternatively called "archeus," "soul," or "vital force." All these various theories assumed that such complicated "cooking" could not take place without a chef, without the ideal pre-existence of special recipes. The approach taken by Iatrophysics and Iatrochemistry to the problem of life was complementary, but the followers of the two schools considered themselves as irreconcilable adversaries. It was only in the eighteenth century that chemical investigations of life began to be considered as part of a materialistic approach that was not in conflict with mechanical explanations. Hermann Boerhaave (1668–1738), the great Dutch scholar, was one of those who proposed an eclectic physical-chemical interpretation of physiological functions and who understood perfectly that there is no conflict, even no fundamental difference, between physics and chemistry.

Three main tendencies can be distinguished in the development of mechanical theories over the course of the eighteenth century. The first one is the spiritualistic interpretation of the organism-machine analogy in Stahl's doctrine of animism. Here, a human being was considered to be composed of an active, intelligent soul and a passive, machine-like body. The Cartesian soul was thinking but was not physiologically active, in contrast with Stahl's *anima*, which was supposed to take part in all living activities. It seems paradoxical that the most vitalist theory of living directing forces, as expressed by the school of Montpellier, was itself based on a mechanical interpretation of vital functions. As one example of this "vitalistic mechanism," we can consider P. J. Barthez's *Nouvelle méchanique des mouvements de l'Homme et des Animaux* (1798).[34]

The second trend of mechanical materialism of the eighteenth century was represented by the followers of Leibniz's philosophy. For them, matter was not something completely passive but was thought to be composed of animated monads, which possessed the powers of perception and motion. These new dynamic views were introduced into biology and medicine by Friedrich Hoffmann (1660–1742), who in an extremely clever way combined physics, chemistry, and biology. Hoffmann gave his doctrine the name of "rational" system.[35] Only the lack of greater knowledge of the facts can account for the weakness of this otherwise logically and very solid general interpretation of organisms.

The third trend was a commitment to consequent and absolute materialism, as, for example, expressed by French philosopher and physician Julien Offroy de La Mettrie (1709–1751). Descartes explained everything mechanically but intelligent action. He stated: "For we can certainly conceive of a machine so constructed that it utters words, and even utters words which correspond to bodily actions causing a change in its organs. . . . But it is not conceivable that such a machine should produce different arrangements of words so as to give an appropriately meaningful answer to whatever is said in its presence, as the dullest of men can do."[36] In his book *L'Homme machine* (1748), La Mettrie reacted against this spiritualistic component of Cartesianism,[37] as he wished to overcome (1) the Aristotelian principle of an immobile God and a God-like soul as the necessary origin of all movement and (2) Descartes's notion of two substances. La Mettrie, thus, endeavored to provide a mechanistic interpretation of the soul. As Vartanian emphasized in the introductory part of his excellent edition of La Mettrie's treatise, "besides the mind-body correlation, another essential feature of *L'Homme machine* is its attempt to prove that the organism as such possesses inherent powers of purposive motion."[38] Actually, La Mettrie's concept of matter is closer to Leibniz's dynamism than to the Cartesian notion of passive *res extensa*. For La Mettrie, all biological structures possess a self-moving power whose most evident expression is muscular irritability: "The human body is a machine which winds its own spring," he commented.[39]

For La Mettrie, man is so complicated a machine that it is impossible to get a correct idea about him *a priori*. Descartes was wrong in reconstructing the human machine only by means of his own thinking. The study of man and animals must start with experience. "Thus it is only *a posteriori* or by trying to disentangle the soul from the organs of the body, so to speak, that one can reach the highest probability concerning man's own nature, even though one cannot discover with certainty what his nature is."[40]

Medical practice taught La Mettrie that feelings and behavior are dependent on drugs, disease, and other material factors. "A mere nothing, a tiny fiber, something that could never be found by the most delicate anatomy, would have made of Erasmus and Fontenelle two idiots."[41] It seems that the starting point of La Mettrie's materialistic interpretation of the school was merely self-observation. According to Frederick the Great's eulogy on La Mettrie,

> During the campaign of Freiburg, La Mettrie had an attack of violent fever. For a philosopher an illness is a school of physiology; he believed that he could clearly see that thought is but a consequence of the organization of the machine, and that the disturbance of the springs has considerable influence on that part of us which the metaphysicians call soul. Filled with these ideas during his convalescence, he boldly bore the torch of experience into the night of metaphysics; he tried to explain by the aid of anatomy the thin texture of understanding, and he found only mechanism where others had supposed an essence superior to matter.[42]

Certainly, the real determination of La Mettrie's thinking was more complicated, as he was impressed by both Newtonian physics and Boerhaave's biochemistry and biophysics.

In the same year La Mettrie published his book, the French Academy of Science admitted to its membership Jacques de Vaucanson (1709–1782), a simple mechanic whose principal claim to fame lay in his construction of automata. Vaucanson constructed a flute player; a duck that could move its wings, swim, eat, and excrete; and an asp that could hiss and dart on Cleopatra's breast. Vaucanson's machines demonstrated or, better, gave the illusion of demonstration, that man could mechanically depict living processes. Certainly, all this now seems very naïve. Today, even a much more complicated robot than Vaucanson's flute player would not be seriously interpreted as an imitation of an organism. Vaucanson's philosophical background is thus hardly understandable for us, and it is surprising to see how significant the theoretical impact of his constructions upon physiologists has been. Recently, Doyon and Liaigre have stressed the historical relationship between medical research and the construction of automata. They showed very well that the principal aim of Vaucanson's efforts was to explain physiological phenomena by an almost perfect mechanical imitation considered not as a simulation but as a real copy of nature's work.[43] Vaucanson called his automata "moving anatomies"; technical achievements that were preceded by a detailed study of anatomy and physiology. His duck (constructed about 1733), the *"canard digérateur,"* was able to digest, that is, to swallow grains and, a little later, to eliminate some excrement-like matter. It was, in fact, a dishonest technical trick and not an imitation of digestion; but for us it is significant that so many people accepted Vaucanson's claims as a real solution to a complex physiological problem.

In 1741, Vaucanson announced to the Academy of Sciences and Arts of Lyon that he had "in mind the project of constructing an automaton which, in its movements, would imitate some important animal functions such as blood circulation, respiration, digestion as well as the working of muscles, tendons, nerves, etc."[44] The construction of a mechanical model of blood circulation was not a new idea. Soon after Harvey's discovery some attempts were made in this direction, especially in Germanic countries (for example, Salomon Reisal's plans for a *"statua humana circulatoria"* in 1674).[45] And Vaucanson was not alone in France either in working on an "artificial man." Claude-Nicolas Le Cat (1700–1768), a famous surgeon, presented to the Academy of Rouen a *Dissertation sur un homme artificiel dans lequel on verrait plusieurs phénomènes de l'homme vivant* (1744).[46] Such a mechanical automaton was supposed to have the value of a philosophical demonstration.

As I pointed out before, all this seems rather naïve. But our models of life phenomena, for example the analogy between the human brain and the computer, are actually not very different, philosophically speaking, from Vaucanson and Le Cat's artificial man. In both cases, some patterns of living phenomena are truly explained but the imitation includes only a part of the biological reality. In both cases life is compared to machines that are not completely realized but only at the very beginning of their technological development.

Returning to La Mettrie, no doubt modern readers find his materialism shockingly simple and lacking in technical details, the author conflating conditions and essence. The dependence of psychic phenomena on material conditions is not sufficient for a materialistic theory of the soul. Curiously enough, La Mettrie was still thinking in terms of Galenic humors, which "according to the nature, abundance, and the different combination . . . make each man different from another."[47] La Mettrie did not give the key to the great enigma but he offered a program for the future in influencing the materialistic philosophy of Helvetius (1715–1771), d'Holbach (1723–1789), and especially the psycho-physiological doctrine of Cabanis (1757–1808). Considered as a program for future research, La Mettrie's treatise is still valuable and in 1928, Joseph Needham (1900–1995), an eminent English embryologist, published a book in which "the several arguments and observation of M. de La Mettrie are carefully considered" and, he argued, "although two hundred years old, are shown to be, in a sense, very justified."[48]

Thomas H. Huxley (1825–1895), one of the most famous biologists of the nineteenth century and a supporter of Darwinism, defined physiology as "the physics and chemistry of life."[49] In fact, the new physical and chemical methods of investigation introduced into biology after the Scientific Revolution in the seventeenth century—and applied to a greater extent in the nineteenth century—have become extremely useful. We are still impressed by the excellent results of physico-chemical analysis of living phenomena. Actually, the greatest progress in biology in the past one hundred years has been due to this approach. The analogy between organism, on the one hand, and machine—or chemical factory—on the other, was a powerful explanatory model.

Is this analogy only an explanatory model, and a very temporary one, or does it mean something more profound? The answer to this question is still open. Even Needham concludes with a positivistic statement, which is that neo-mechanism "recognizes the supreme jurisdiction of the mechanistic theory of life, but admits it at the same time to be a methodological fiction."[50]

PART THREE

History of Science:
The Laboratory of Epistemology

A Plea for Freeing the History
of Scientific Discoveries from Myth

In the course of the historical reconstruction of a scientific discovery, a process occurs similar to what Stendhal, who used this image to study the hazards of love, called crystallization: "At the salt mines of Salzburg a branch stripped of its leaves by winter is thrown into the abandoned depths of the mine; taken out two or three months later it is covered with brilliant crystals; the smallest twigs, those no stouter than the leg of a sparrow, are arrayed with an infinity of sparkling, dazzling diamonds; it is impossible to recognize the original branch. I call crystallization that process of the mind which, from everything which is presented to it, draws the conclusion that there are new perfections in the objects of its love."[1]

Our "objects," for us, historians of science, are the whole of *res gestae* (lived history) and *res scriptae* (written history) of past scholars. We embellish these in our fashion and we contrive to give them a luster that is in harmony with our general ideas on the nature of knowledge and on the way by which it is acquired. So, the dry branch of "historical facts" is, by a process of progressive theorization, enveloped little by little in a thick layer of seductive "myths."

Let us take for granted that it is utopian to wish to produce, in accordance with Ranke's precept (or, to keep within the more limited field of the history of science, with Sarton's positivist recommendations), a history claiming to be the absolutely objective mirror of what really happened.[2] Let us also take for granted that historical reconstruction carries out selections, makes connections, and proposes "explanations" that necessitate a philosophical stand, indeed, an involvement of an ideological nature. But this recognition of a certain lack of objectivity in historical research should urge us not to relativism and an attitude of resignation but to a careful and systematic analysis of the factors that intervene in "crystallization," to the exposure of the processes that condition our search for truth. It seems inevitable nowadays that we demand that the historian of science submit himself to an epistemological self-examination. In short, at the moment when he becomes fully conscious of the impossibility of total objectivity in his discourse, the historian should feel more than ever the necessity to elucidate, as much as possible, the modalities of his deviations therefrom.

Certain methodological presuppositions appear so inherent in scientific thought that they cannot be questioned without jeopardizing the rationality of discourse. But there do exist key ideas as well that are less untouchable. Historians of science often draw their inspiration from convictions whose validity is not certain: I propose to call these convictions *methodological myths*. These "myths" have historical roots and fulfill an important psychological and social role. It is a question, fundamentally, of beliefs and not of proven assertions, nor of axioms indispensable for all subsequent research. These "myths" are rationalized justifications of our desires. I therefore refer to them also as *methodological illusions*. Their great strength is derived from the fact that we are indeed dealing here with "truths," but alas, with only partial truths. These myths consist most often, to put it concisely, in the extension of an idea beyond its range of validity by transforming a partially sound rule of limited application into an illusory general rule.

The historical reconstruction of scientific discovery is dependent on a particularly rich and tenacious "mythology." Here are a few examples of these governing ideas, which, diverse as they are in character and in their epistemological level, are both widely accepted and of doubtful validity. The following list is not systematic and has no claim to exhaustiveness. Its sole aim is to draw the attention of historians of science to the existence of the hidden traps strewn in the path of the historical rediscovery of scientific discoveries.

1. The Myth of the Epistemological Adequacy of "Anatomizing"
Acquired Scientific Knowledge and the Patterns of Its Conquest

Many epistemologists are interested exclusively in the analysis of the structure of scientific knowledge and put little store by the study of the historical coming into place of each structural element of present knowledge. Most neopositivists belong to this category: They conjure away the problem of the actual genesis of the network of scientific propositions by deliberately substituting its formal logical construction for its actual historical becoming.

It is quite significant that Israel Scheffler (1923–2014), a philosopher who teaches the logic of scientific investigation at Harvard University, entitled his main work *The Anatomy of Inquiry*.[3] Indeed, anatomy, a morphological and static science par excellence, is concerned with the state of structures and not with the process of their formation. Undoubtedly, the analogy with the anatomy is inescapable to characterize the preoccupations and working methods of certain philosophers of science.

However, nowadays one hears more and more the voice of those who are interested not only in constituted science, in what seems immutable in the process of scientific research, but also and especially in science-in-the-making, in the genesis of so-called scientific structures. By widening the field of epistemological reflection, they complete "anatomical" analysis with a kind of "embryology" of scientific knowledge.

In this last case, it is relatively easy to insist on the necessity of recourse to historical method. But I feel one cannot altogether do without it even in an epistemological study that aims at probing into the subtle details of established science. The analogy between the morphology of organized bodies and the state of scientific knowledge appears unsatisfactory in this connection: The structural relationships of scientific thought cannot be represented adequately or studied in the full richness of their content if their manifold historical articulations are not taken into account.

As for those epistemologists concerned with the genesis of scientific knowledge, one can aim the following criticism at them: They often offer explanations of the process of scientific discovery when its different stages are insufficiently known and described. In short—keeping within the limits of Scheffler's analogy—it appears that a "causal embryology" is made without a previous grounding in the "descriptive embryology" of the whole range of scientific matters.

In another paper,[4] I have argued that the philosophical theories concealing the development of science have a scientific content only insofar as they can be compared with the data of the history of science. This is a fundamental requirement with which philosophers must comply. I would now like to emphasize that it puts historians of science in a delicate position, for it assigns a task to them that is still far from accomplished and raises questions to which one must beware of replying hastily or in an oversimplified or complacent manner. That brings us to the next section.

2. The Illusion Concerning the Adequacy of Our Present Historical Knowledge for Reconstructing the Process of Scientific Discovery or for Validating Hypotheses Relating to This Endeavor

One is nowadays aware, on the one hand, of an extraordinary proliferation of publications of a general nature on the creativity and the intrinsic mechanism of discovery processes and, on the other hand, of an appalling dearth of solidly documented historical studies of concrete discoveries. This dearth is not due to a shortage of writings of a biographical nature on the researchers and their scientific works—in absolute numbers there are actually too many. But they untiringly repeat the same anecdotes and are satisfied with a superficial documentation that is not sifted with critical appraisal and is dressed up to comply with the demands of a new kind of hagiography. One of the main weaknesses of present-day publications on scientific discoveries is that the level of abstraction and generalization is too high, thus creating false problems and masking certain essential aspects.

The fact is that epistemologists have nowadays at their disposal only a few precise and detailed descriptions of historical events relating to important discoveries. That is why general theories on scientific creativity do not inspire the experienced scholar with confidence and can proliferate with impunity for the indulgence of philosophical speculations.

L. Pearce Williams (1927–2015) notes that epistemologists often base their theories concerning the genesis of scientific knowledge either—as Kuhn does, for example—on "what scientists *do*" (without, however, providing the irrefutable historical proof that they really act this way), or—as in Popper's case—"on what they *ought* to do" (with a historical documentation that is too interpretative and restricted). One would rightly wish to base these epistemological systems on the history of science but "it cannot bear such a load at this time."[5]

Williams's critical remark has been accepted by Kuhn (1922–1996) but rejected by Popper (1902–1994), who places logic outside and beyond the bounds of history. Certainly, it is reasonable enough to admit that logic has nothing to learn from history. But this is not the problem, which lies in the possibility, or rather the great likelihood, that the structure of science and, even more, the process of its growth do not belong entirely to the field of traditional logic.

Certain attempts at a historical revision of "classic cases" (let me quote as example, in the one field of the life sciences, the work of W. Pagel [1898–1983] on Harvey, that of H. E. Gruber [1922–2005] on Darwin, and that of F. L. Holmes [1932–2003] and my own on Bernard)[6] show that a new light can be thrown on subjects one might have thought to be exhausted. In the history of scientific discovery, all or almost all is yet to be done over again. Historians sensitive to the problems of modern epistemology must systematically re-examine the past course of science. Alas, we shall be disappointed too often: Historical sources at our disposal rarely measure up to our wishes. It would be illusory to draw a supposedly decisive epistemological lesson from what we can know with certainty about the discoveries of Aristotle (384–322 BC) or Galen (129–217/216), even Copernicus (1473–1543) or Vesalius (1514–1564). We must therefore concentrate on a few instances privileged by the nature and abundance of the surviving documentation. But as we expose the shortcomings of our historical sources, would it not be possible to improve our methods of collecting and preserving present-day evidence of scientific discovery, thus to prepare and render more fruitful the work of future historians?

3. The Myth of a Perfect Agreement between a Rational Reconstruction and the Experience of Discovery

A. DEFORMATION OF REALITY IN THE RESEARCHER'S REPORT

The fundamental sources on which any historical reconstruction of a scientific discovery must draw are documents originating from researchers directly involved in the different stages of its realization. These sources, whatever their nature (publications, correspondence, notebooks, laboratory records, recorded statements, recollections of conversations by a third party, etc.), can provide reliable information about the workings of someone's mind and his conscious motivations at a given moment or, at least, express irrevocably what that person wished to record for himself or convey to his contemporaries or to posterity.

We are here in touch with first rate "historical facts" that we must do our utmost to garner, the contents of which, with no exception, must be taken into account. But these documents do not in themselves convey the picture of the process that concerns the historian of science. He has to interpret them, build an account that, via the sum total of available historical traces, offers a rational reconstruction of that historical order the existence of which we take for granted in what I called "*le vécu de la découverte*" ("the lived experience of discovery")—that is, in what really happened in the course of the actualization of a significant growth of scientific knowledge.[7]

In this reconstruction, one runs a great risk of going right off the track in always accepting literally what the protagonists or witnesses of an event have to say. Testimony, direct or indirect, is a kind of raw datum that must be used beyond its immediate and superficial significance. A confrontation of different sources is called for to avoid the traps of accounts that are too subjective, made in good faith, and yet often false because of their partiality.

An illusion, both widespread and deeply rooted for reasons that are easily understandable, consists in the belief that perfect reconstructions of the experience of discovery can be found implicitly in the writings of the researchers concerned. Is not every individual, if not the best judge, at least the one most acquainted with his personal history, his own life experience? This widespread belief and popular opinion appears highly questionable to me. With regard to events that have a heavy emotional charge and important social implications for the person in question—which is certainly the case of most scientific discoveries—an autobiographical account is certainly valuable but cannot be accepted as definitive.

Scientific works swarm with surveys of the history of the research and discoveries reported therein. Their authors often present in them the order of their reasoning and of their experiments, acknowledging the influence exerted on them by the knowledge of their predecessors' works and opinions, etc. However, such information has no really historical purpose. Here we find not history as such, but elements for a future history. Addressing the scientific community to inform it about the state of their researchers, scientists provide certain details of their personal histories so as to explain better the content of their discoveries, not to analyze the actual process of their attainment.

A few great scholars have published, as an afterthought, autobiographical texts reconstructing the progress of their reasoning and even containing, as a by-product, reflections on scientific creativity. To name a few at random (still taking examples only from the life sciences): Claude Bernard (1813–1878), Hermann von Helmholtz (1821–1894), Otto Loewi (1873–1961), Charles Nicolle (1866–1936), Karl von Frisch (1886–1982), Walter B. Cannon (1871–1945), Albert Szent-Györgyi (1893–1986), Hans Selye (1907–1982), John Eccles (1903–1997), James D. Watson (b. 1928), Jacques Monod (1910–1976), etc. Their contributions to the history of science and epistemology are invaluable. No one would dare to deny the importance of their testimony on the "facts" and their analysis of historical connections regarding their own self. Nevertheless, historians cannot be content simply with transcribing these individual presentations and incorporating them in the more general framework of a treatise of the history of science. Let us take a few examples.[8]

(a) In order to be aware of the difference of method and perspective that separate the autobiographical account from the discourse of the historian of science, it is useful to compare the writings of Watson and Olby who retrace, each in its own fashion, the path leading to the discovery of the double helix of DNA.[9]

(b) With the help of detailed studies of Claude Bernard's research (in particular that concerning the glycogenic function of the liver, the so-called *piqure sucrée*, the vasoconstrictor nerves, and the mechanisms of intoxication and drug-induced reactions), I have tried to show the points of disagreement that exist between the lived experience of a discovery and its verbal or written presentation.[10] The discoverer's account is an abridgement of the experience he has lived; yet it involves not only a simplification through omission of the unimportant and secondary but also a rationalizing readjustment, whereby the actual sequence of events is transmuted in favor of logical rigor and coherence. A winding path becomes a straight-lined road, with sudden and precise changes of direction.

(c) Galileo (1564–1642) maintains that he invented the telescope as a result of precise optical reasoning. Now, as Ludovico Geymonat (1908–1991) has clearly shown, in 1609, at the time of his invention, Galileo did not have the theoretical knowledge necessary for the logical argument he claims in his belated reminiscences.[11]

(d) Analyzing Loewi's account of his discovery of "Vagusstoff,"[12] Zénon Bacq (1903–1983) points out that this reconstruction of events is too elegant and, in some degree, inevitably false, for Loewi had forgotten that at the moment of his discovery, his technique of electrical stimulation of an isolated frog heart was not yet able to separate completely the inhibitory cardiac nerves (vagus fibers) from the sympathetic fibers.[13]

(e) The same internal impossibility can be found in Claude Bernard's famous account of his discovery of glycosuria provoked by a lesion in a well-defined area of the central nervous system. He quotes as a theoretical starting point for his decisive experiment an analogy with certain features of the neural influence on saliva secretion that were then, in truth, totally unknown to him.

I should like to stress the interest presented by historical investigations into the progressive transformations of personal experience into a message addressed to others and of the original information into a fairy tale intended to amaze the scientist himself as well as his peers and the general public. A meticulous comparison of documents bearing different dates but coming from the same person and recounting the same event, as well as psychophysiological studies on memory and sociopsychological ones on testimony, will perhaps allow certain constant features to emerge in the distortion of historical reality made a posteriori by the protagonists themselves.

In my work on Claude Bernard's experimental reasoning, I took a few tentative steps in this direction. I have been able to establish, for example, that in the lived experience of his work, the researcher wishes to convince himself, while in his report, he wishes to convince others. As a result, the real starting point of the discovery, often too fragile, finds itself reduced more and more to a secondary argument, to an unimportant and even irksome entity. Besides, the most common error of the autobiographical report is chronological inversion in the sequence of events or the concatenation of ideas.

Be it said in passing that sometimes one comes up against what I have called the "illusion of narratives' time dimension." What is recounted no longer takes place in real time since the end conditions the beginning. A discovery is achieved in a real area of time in total ignorance of the future, while the researcher's report and the historian's discourse are made on an artificial time-fabric that resembles real time in its structuration but that differs therefrom in that is has no becoming—that is, it forms a closed whole.

B. THE DEFORMATION OF REALITY BY THE HISTORIAN
OR THE PHILOSOPHER OF SCIENCE

This deformation may be voluntary or involuntary; in a large number of cases it is betwixt and between. The most typical is the distortion that takes place without a conscious decision to cheat but nevertheless with the intent of proving, or rather illustrating by an historical example, a preconceived general idea. Abel Rey (1873–1940), for example, has shown in a polemic against Duhem's conventionalism how much a particular philosophical attitude leads the latter to separate artificially the presentation of scientific "truth" from the real process of his discovery.[14]

For an observer who places himself a little above the present epistemological fray, however, it is evident that the logical reconstruction of the way in which a science (for example, classical mechanics, quantum physics, or molecular biology) is constituted, and further still the way in which a new paradigm has asserted itself in a defined field of scientific research, does not tally, and does not in fact aim at tallying, with absolute objectivity, with the actual historical process. On the occasion of another seminar,[15] I have tried to prove that each historical example can be harnessed to any reasonable epistemological position whatever, provided selection, with a given slant, of the historical documentation is made and certain details are declared unimportant.[16]

The reasons for the distortion with which we deal here are most often of an ideological nature. It springs from an idea anticipating the historical inquiry. This is not at all surprising. The "patterns of expectation" are already at work at the level of the organization of perception and their interference is all the more to be feared as one climbs the ladder of theorization.

The advent of awareness vis-à-vis the myth of the absolute objectivity of historical discourse should not give rise to a facile and baneful attitude of historical relativism and the triumph of skepticism. Criticism is the only healthy reaction. Distortions do exist, but they can be detected by critical analysis. The neutrality of the historian and the philosopher is never absolute in practice, a fact that does not, however, exclude the possibility of acquiring, little by little, an increasingly objective knowledge. Like scientific research (in the narrow sense of the word), historical inquiry is susceptible to progressive objectivation, to an asymptotical approximation to a reality independent of the researcher's mind.

4. The Myth of an Impersonal Historical Development of "Scientific Thought"

Under this heading I wish to denounce both (a) the pseudo-problem of the conflict between "internal" history and history referred to as "external" and (b) the illusion resulting from the confusion between the personal history of the scientist and the general history of the progress of scientific knowledge.

Without doubt, a dual aspect exists of the *itinerarium mentis ad veritatem*:[17] on the one hand, the actual itinerary of ideas in the "discoverer's mind," and on the other, the historical development of "scientific thought" considered in abstracto—that is to say, without taking into account the particularities of those individuals who are unavoidably its mainstay.

Experience proves the possibility of a historic discourse on the transformation of scientific theories that disregards considerations of a biographical or psychological nature on those who have in effect created that world of ideas. Is such a discourse legitimate? Although I am convinced that thought is exclusively the product of concrete and individual nervous systems, I admit the pragmatic value of studying an "internal dynamics" of scientific ideas. However, to overcome the paradox of a thought without a head to think it, it appears necessary to call upon an intermediary notion: that of "scientific community." I therefore define "scientific thought" as a structured aggregate of known facts and propositions of a certain type that are accepted or debated by a group restricted in historical time and in social space. In this way, discourse on the general development of scientific thought relates to a historical reality and not to an idealistic schema.

Evidently, if the "internalist" approach is conceived, justified, and realized with the help of the notion of "scientific community," it concerns, strictly speaking, not the "general" development of scientific ideas but only numerous sectors of it. The success at the monograph level of such a historical presentation will provide proof of the reality of a community of scholars within well-defined spatiotemporal coordinates, with due regard to the problems they study. "Internalist" scientific historiography requires a complementary inquiry: the historical examination of the community concerned—that is to say, an "externalist" analysis that is essentially of a sociological nature.

In their common wish to be mutually exclusive, approaches known as "internal" and "external" become myths. In fact, they are self-limiting, interdependent, and mutually complementary. Their conflict is nothing but a false problem, born of exaggerated pretensions on both sides. Even someone interested only in the history of a specific scientific discovery, who would like to isolate it from the texture of general history, must have an overall picture of the flux of scientific ideas, as well as take into account economic, political, and sociopsychological factors. But we must never forget that this overall picture refers to a second-degree reality born of a theoretical construction.

The danger lies in the errors of fact and interpretation that can occur if this "history of scientific thought" is projected into the consciousness of a scientist qua individual. That, alas, happens too often. To study properly the creative process of new ideas, it is not enough to know what, at a given moment, was "known" on the topic under consideration, but one must try hard to state precisely what each scholar involved in the matter knew or believed he knew. In my work on Claude Bernard's experimental reasoning, I have wished to show clearly that it is necessary to distinguish carefully between the two types of general reconstruction of ideas on intoxication by carbon monoxide and a reconstruction of the particular evolution of known facts on this subject in a researcher's consciousness.[18] These two reconstructions overlap only partially.

An additional task, therefore, devolves on the historian of the process of scientific discovery: He must establish if, at critical moments, the "discoverer" has really known (and, if so, how and in what precise form) certain "facts" and certain "explanations" prior to the discovery and well known to the scientific community. To use the example quoted above, let us remember that Claude Bernard discovered, on his own, the "arterial" color of venous blood poisoned by carbon monoxide. This observation was previously described by other scientists, but Bernard was not aware of their work; that he achieved it by his own personal effort greatly influenced his subsequent research and facilitated the formation of new ideas.

Let us mention in this context the sometimes-positive role of ignorance of an idea by the discoverer of a better one: In such a case, the former idea was an "epistemological obstacle" for the other researchers.[19] Some scientific discoveries and technical inventions have been made because their author did not realize the "impossible nature" of this endeavor.

5. The Myth of the Unity of the Discoverer, of the Place and Time of the Discovery

Most important scientific discoveries have not been achieved by a single person, and even less at a precise moment in such a person's life. However, in many historical accounts we find in this connection variations of the old myth of the birth of Pallas Athena, the embodiment of knowledge, springing fully armed from the head of Zeus. This is a false image: Discovery does not spring in one leap, radiating beauty and strength, from the creator's head.

To define a scientific discovery, a brutal intervention into a complex spatiotemporal reality and the isolation of certain events with a near-symbolic value that encompasses not just one point but a whole field of action is often necessary. That makes schematization inevitable, but this can easily become abusive, for one goes almost imperceptibly from a simplification to a historical illusion.

Asking the question "Who discovered this (a substance, a relationship, etc.) and at what time?" assumes an improbably simple train of events, a particular historical situation the existence of which is sometimes impossible to find at the end of an inquiry and that one is not allowed to take straightaway for granted. Kuhn[20] has shown the misleading nature of such a question with regard to the discovery of oxygen. I reached the same conclusion about the discoverer and the moment of discovery of the mechanism of curarization.[21] The desire to reply at all costs with a name and an exact date is, in the cases quoted, tantamount to weakening historical reality in a way that modifies and betrays its true meaning.

Let us examine, as an example, the following question: "Who discovered glycogen and when?" On close inspection, we notice that this question, of the kind found in scientific textbooks (and to which all good encyclopedias give a reply as crisp as it is problematic), is precise only in appearance. What is the meaning of "to discover glycogen"? Does it mean to be aware of its existence? To give it a name? To isolate it (and, in this case, how purely)? Or to know its chemical composition? Claude Bernard was the first to have an inkling of this substance. Following an original interpretation of certain experiments, he assumed its existence (1855); he called it "glycogen matter," then simply "glycogen," thus defining its principal functional property. He also foresaw the possibility of its extraction from the liver. However, Claude Bernard was not the first to isolate glycogen. He was preceded, in July 1856, by a German student, Christian Andreas Victor Hensen (1835–1924), who achieved this feat thanks to the advice of his professor, Johann Joseph Scherer (1814–1869), inspired in turn by Bernard's publications. Was this simply a practical application of Claude Bernard's "theoretical discovery"? To give credit to Hensen's work, one must acknowledge that Bernard's hypothesis was still quite uncertain, that the procedure employed by the young German scholar was original, and that, in a way, he helped Bernard to perfect his own method of isolating glycogen. Only in March 1857 did Claude Bernard obtain a relatively pure glycogen. Physiologist E. F. W. Pflüger (1829–1910) showed by repeating Hensen's experiments that the substance obtained by the latter in 1856 contained more proteins than polymerized carbohydrates. If it is acceptable to look upon Bernard as "the" discoverer of glycogen, it is only on condition that we do not forget that he succeeded in this research because he was immersed in the efforts of a scientific community. But it is illusory to try to assign a precise date to what was, in fact, quite a long process, made up of many stages.

Replies to questions of the kind described above are "true" only with regard to a certain level of historical explanation. Paradoxically, their "veracity" can be inverted when this level is changed.

For example, it is said that Claude Bernard discovered hepatic glycogenesis. Taken literally, this assertion is false: Bernard was even aggressively opposed to contemporary scientists who believed in the existence of "hepatic glycogenesis" (in the present sense of the term, namely the storing in the liver, in the form of glycogen, of sugar coming from food ingestion). His "glycogenesis" was glucidic synthesis derived from other chemical substances ("gluconeogenesis" for present-day physiologists). Louis Figuier (1819–1894), an opponent of Bernard's opinions on this topic, was often right in the details, but it is nevertheless fair to attribute to Bernard almost all the glory of the discovery of the hepatic metabolism of sugars. The statement quoted at the start of this paragraph is both false and true: It becomes valid in a meaning that corresponds to a higher level of abstraction in the presentation of the history of ideas.

6. The Positivist Myth of the Straight Road to Truth

Neither the meandering of individual thought nor the advances of scientific knowledge within a community proceed by successive approximations, always in the right direction, toward truth. The path to discovery is a winding one.

Positivist historiography has accustomed us to representations of the development of scientific knowledge that are like stairways rising triumphantly toward the temple of present-day science and made up of steps each of which represents a "positive" scientific advance, a partial but definitive truth. This myth still predominates today in most "historical introductions" found at the beginning of chapters in textbooks of contemporary science. This is a Manichean illusion where good and evil, truth and error, are clearly separated. This triumphant and didactic vision of the history of science played an important part in the social assertion of scientism in the last century, but it does not stand up to a thorough examination of historic reality.

The ascent toward truth is not linear, and error is not necessarily (though often it is the case) a negative element—that is, an obstacle on the road of scientific research. Some "errors" are particularly prolific, more functional, in a given historical situation, than opinions recognized as more "true" in the judgment of posterity. Thus, historical investigations carried out recently or still going on amply demonstrate that the influence of magic and alchemy on the birth of the new sciences in the seventeenth century was not altogether a negative one as positivist historians of the last century would have us believe. Even men like Francis Bacon (1561–1626) and Newton (1643–1727) are indebted to the hermetic tradition and to ways of thinking condemned without appeal by a certain brand of rationalism.

A scholar's judgment on an opinion considered as a scientific image claiming to give the truthful explanation of the world should not be conflated with the historian's judgment of this same opinion considered as a guiding idea (*idée-force*) influencing the subsequent development of scientific research.

For a historian blinded by an uncompromising positivism, the fact that Miguel Serveto (1511–1553) included in a theological work his description of the passage of blood through the lungs can only be an accident with no special significance or, at best, the testimony of a man torn between the light of reason and the darkness of his chimerical speculations. Yet, looking at it closely, in this case as in so many others, the light of discovery shone forth owing to a piece of confused and analogical reasoning wherein mysticism was an indispensable component. A particularly striking example of the mixture of "bad method" and "good results" is found in Kepler's work. A modern scholar has gone so far as to say that this genius would appear much greater to us if three quarters of his writings could be excised.[22] The "normative" aspect of this opinion renders it odious and dangerous. Kepler, Newton, and Harvey did not make their discoveries in spite of their astrological, alchemical, or Aristotelian prejudices but through them and in part because of them. About the work of Kepler, Catherine Chevalley notes very aptly that "a history of science which limits itself to tracing the skyline of the world of geniuses prevents itself from understanding the mechanism of scientific research; conversely, the genuinely philosophical task is not to locate truth and to condemn error, but to understand the intellectual processes which are now alien to us."[23]

The discovery of the circulation of the blood has often been cited as a typical example of the rigorous and successful application either of Baconian induction or of the hypothetico-deductive method. These approved interpretations, dear to the particular inclination of every positivist historian, are not altogether consonant with what the documents say about the real genesis of William Harvey's discovery. In this respect, we may not disregard the influence of the mystical symbolism of circles, Gnostic philosophy, and Peripatetic metaphysics. In physics, even, Harvey drew his inspiration as much from Aristotle as from Galileo's new mechanics. When he gives the heart the distinguished title of "sun of their microcosm,"[24] Harvey uses something more than a poetic image. Physiological cardiocentrism is actually historically bound up with astronomical heliocentrism. The most decisive step in Harvey's reasoning was the transformation of the Aristotelian analogy between irrigation of a garden and the distribution of blood in the body, a transformation carried out with the help of another "model" likewise taken from his reading of Aristotle—namely, the place of the sun in the circular motion of terrestrial and atmospheric waters.

For a considerable number of discoveries, the *post hoc* rationalization of intellectual processes involved give the impression of a linear, flawless progress. But unless one can rely on a vast and firsthand historical documentation, it is possible, in virtually all cases, to break down the structure of the official representation of the discovery and to prove the existence of significant errors, aborted attempts, and strokes of luck. I will not dwell on the role of chance here, for I have recently published a study specially devoted to this topic.[25]

To illustrate how the rational reconstruction of discoveries serves as a model for a positivist history, let us remember the introduction of vaccination against smallpox by Jenner (Razzell[26] has shown certain of its "mythical" aspects), the first vaccinations against rabies by Pasteur, and Bernard's experiments on glycosuria. The events leading to these achievements are not in reality as simply and logically linked up as we are asked to believe.

A particularly instructive example is provided by the history of the discovery of insulin. It is admirably clear in the justification of the Nobel Prize award by the chairman of the selection committee, and in the autobiographical accounts of the recipients, but becomes more and more confused as one proceeds in the critical analysis of the facts. In the first place, it does not seem permissible to eliminate from the historical report the contribution of certain "forerunners" (see Murray's article on experiments by Paulesco, not unknown to, but willfully deformed by, the winning team).[27] What is even more interesting, from our point of view, is that one becomes aware (see especially the excellent study of Pratt)[28] that the official version of the way the discovery was arrived at does not stand up to an examination of the original publications or to a repetition of the experiments. The clue of Banting's reasoning was simply an illusion: Obstruction of the pancreatic ducts does not give rise to a selective degeneration of the tissues producing trypsin. Contrary to the working hypothesis he proposes, the extract obtained from a normal pancreas contains a higher proportion of insulin than that yielded by a "degenerate" one. The results of Banting's and Best's experiments are at variance with the conclusions that they drew from them. We know today that the success of the Toronto team was due to the precipitation of proteinic hormone by 95 percent alcohol, which eliminates toxic ingredients and allows medical use of the extract. The inventor of this process, the biochemist James B. Collip (1892–1965), was put on the side, when the Nobel Prize was awarded, to make room for the head of the laboratory in which the discovery had been made.

7. The Myth of Continuous Evolution and the Myth of Permanent Revolution

Continuity and rupture coexist and any historical venture based on an exclusive mythicizing of one of these two complementary aspects would seem doomed in advance. If, in the personal history of a researcher, there are sudden moments of enlightenment, comparable to Archimedes's *Eureka*, there is also, inevitably, prior to them an incubation period, a slow, continuous process of intellectual ripening. In the general history of the sciences, there are likewise revolutionary reversals—historical falsifications in Popper's sense of the expression or paradigmatic reversals in Kuhn's terminology—but as the latter so well emphasizes, that does not exclude the existence of a slow, not really discontinuous accumulation of scientific knowledge in the period separating two paradigmatic crises.

From this standpoint let us look at the history of the discovery of the circulation of the blood. Baldini has examined Harvey's work in the light of Popper's epistemology and has severely criticized the idea that his discovery resulted from a progressive and gradual piecing-together of a mosaic of minor discoveries, of tiny steps made by his "forerunners" and by Harvey himself, necessarily leading by their added momentum to the new pattern of circulatory physiology.[29] Baldini's analysis seems convincing: Harvey's achievement is revolutionary and, in essence, cannot be explained by the summation of a series of partial discoveries. It is nonetheless true that such a summation exists as well, and if it does not necessarily produce or explain the decisive discovery, it is nevertheless, for its production, an indispensable precondition. The advent of Harvey's new physiology is unthinkable without the cumulative contribution of Vesalius, Colombo, Cesalpino, Fabrizio d'Acquapendente, and several other researchers who breached the Galenic system and allowed Harvey to find himself confronted with a particular panorama of known facts.

The "heroic view" of all scientific research idealizes historical reality. Without doubt, a "normal science" (in Kuhn's sense) exists, and I cannot participate, in this connection, in the disillusioned sigh ("Yes, alas!") of certain of Popper's disciples. The existence and proper working of this "normal science" is a historical *sine qua non* precondition of "extraordinary science." The alternation of "puzzle-solving" periods and emergence of new scientific theories seems to belong to the very nature of the process that produces the growth of our scientific knowledge.

Certain epistemologists, for example Hanson and Kuhn, compare the act of scientific discovery to the shift in the interpretation of an ambiguous drawing (the famous rabbit-duck).[30] According to this analogy, a scientific theory would explain a group of facts exactly as a perceptual interpretation organizes lines and spots into a gestalt. We would thus pass from one interpretation to another, both in the change of paradigm and in the setting-up of a new apperception, by a total transformation determined by *insight* and therefore of an essentially unforeseeable and radically discontinuous nature. The reversal of the scientific paradigm has been compared to a kind of mystic crisis.

Without wishing to go as far as the "anarchical" excess of Feyerabend, who speaks of the "incommensurability of theories"[31] and the poetic determination of the choices made by scholars, I am ready to accept the existence, in the progress of knowledge, of qualitative jumps—that is to say, changes that are irreducible to the accumulated mass of variations of secondary importance. But I do not wish to fall into the trap of the myth of a permanent revolution and deny the fundamental importance of slow transformations of scientific theories, which come about by a progressive adaptation, not by a reversal of paradigms. The two aspects appear to me complementary.

To go back to the analogy between scientific discovery and the interpretation of the rabbit-duck drawing, I intend to consider it in a more dynamic way than that in which it is usually described. Let us imagine that the drawing in question is not a state but a becoming—that is to say, that it is not static but mobile, in slow, continuous transformations. This can be realized, for example, on the screen of an analogical computer. Let us imagine further that this machine is programmed so that the drawing presented to the observer at the outset suggests beyond doubt the outline of a rabbit and that the contours and spots gradually change to present only at a determinate moment the ambiguous drawing. It is then and only then that the gestalt reorganization of apperception can take place. It can but is not bound to. The gestalt switch depends certainly on the observer's state of mind. To him will be attributed the revolution in the interpretation of the whole, but it must not be forgotten all the same that it was impossible without the preceding stage that gradually, by a cumulative effect, gives the drawing its ambiguity. To bring my analogy closer to reality, one must also imagine that the new apperception reverberates by a feedback on the objective characteristics of the drawing, reinforcing the shape of the duck. The gestalt model of scientific discovery can be made valid only by supposing that each of its parts, the observer's state of mind and the drawing observer, are in a condition of flux and that their link-up works in both directions.

In my opinion, the acquisition of knowledge bears the essential characteristics of *biological growth* in general. The latter is not simply an increase by progressive accumulation of elements; it is always a morphogenesis that is brought about by the setting up of a structure and by the alternation of cumulative transformation (minor spells of imbalance with a continuous regulating mechanism) and of "restructurations" (sudden breaks, accompanied by new levels of equilibrium).

Piaget and Lorenz, the former starting from the observation of children and the latter from an analysis of animal behavior, reached fairly similar conclusions as to the existence of isomorphisms between cognitive processes and biological regulations. According to Piaget's genetic epistemology, the intellectual development of the child passes through progressive "restructurations," through conquest of stages (the morphogenesis of knowledge thus resembles embryogenesis). The psychologist has the impression that progress is made through rapid mutations, but—says Piaget—the fundamental transformation is slow; what is sudden is not the process of building up but the coming of awareness, the eventual comprehension at the moment of completion of the structuring stage.[32]

Claude Bernard has given a similar description of the process of discovery: "Illumination" is only the sudden completion of a slow and "underground" maturation process (the term *subconscious* was not yet fashionable) marked by the assimilation of new data and the construction of explanatory patterns.

Modern mathematization, based largely on the notion of function and on differential and integral calculus, favors an understanding of the continuous, scalable aspect of each event. But a new mathematics of discontinuous entities is now emerging: René Thom's "theory of catastrophes." Born of topological considerations, this outline of a general theory of qualitative models seeks to formulate mathematical characterization of "catastrophes" (that is, sudden reorganizations of states of equilibrium).[33] A large number of biological phenomena belong to this category and I think we must also include the processes of artistic and scientific creativity.

8. An Illusion That Results from the Projection of Initial Circumstances in a Developed Sequence of Events

This illusion is well known thanks to the chicken-or-egg priority sophism. It continues, however, doggedly to live on a form that is relevant precisely to experimental reasoning and the genesis of discovery: Are "facts" (or, rather, the observation of the facts) anterior to the hypotheses or, on the contrary, do hypotheses always precede "facts"? In other words, does experimental reasoning begin with observation or theory?

As in the case of the chicken-or-egg question, it is not difficult to show that each of two sequential elements must by turn precede the other. On the one hand, observation is never, in the actual circumstances of scientific research, independent of a certain theoretical framework, of a network of hypotheses formulated beforehand. I shall return to this with regard to induction. Let us simply recall here that, according to Piaget's investigations, theoretical structuration, already in the child, modulates very strongly the "facts" observed (let us quote, as an example, the experiment with the drawing of the line of level of a liquid in a tilted glass). On the other hand, all scientific hypotheses are, in one way or another, induced or influenced by previous empirical knowledge.

In short, we need theories to observe facts and we need observed facts to produce theories. In actual scientific research, the dialectical interplay between theory and practice, between epistemic structures and empirical content, is already so far advanced that the question of the chronological primacy of observation or hypothesis is improperly formulated and leads to a number of paradoxes.

To follow another line of argument, it may well be that this interplay of influences is not only at a very advanced state but also more subtle and complex than the simple alternation of the two abovementioned elements, an alternation suggested by the generalizations of the hypothetico-deductive method.

9. The Myth of the Clear-Cut and Absolute Alternation
of the Observation of Facts and the Invention of Hypotheses

According to a common opinion, quite widespread among scientists, experimental reasoning passes through three stages, which Claude Bernard had represented by the following schema: $O \rightarrow H \rightarrow E$ (O = observation; H = hypothesis; E = experiment). Observation gives rise to hypothesis, which is used in the carrying out of an experiment offering in its turn a new observation that confirms or invalidates the first hypothesis. Bernard's triad extends and branches out into several chains of "experimental reasoning," for each observation can give rise to hypotheses and each hypothesis can be the starting point for new experiments. Strictly speaking, these chains are made up of two constituent elements and not three, for observation and experiment differ only in their position in the experimental cycle. Experiment is an observation engineered to test the validity of a hypothesis: It is a terminal point with regard to the latter and also, by providing new "facts" for the researcher's consideration, a starting point with regard to subsequent hypotheses. Observation and experiment are enshrined in the empirical tradition, but between them hypothesis is thrown as a bridge: hypothesis, that rational element through which—says Bernard—"the mind of a scientist is always placed, as it were, between two observations: one which serves as starting point for reasoning, and the other which serves as conclusion."[34] What seems essential in this schema is the on-and-off alteration of theory and practice, of imagination and "facts," of the intellectual variations of ideas and their natural selection.

What mental operations are represented by the two arrows of the triad? The second signifies that starting from a hypothesis one arrives by logical deduction at the construction of situations that can be subjected to experimental control. It is the graphic expression of a methodology advocated by men like Galileo and Newton. The crucial problem is to explain the birth of hypothesis from observed facts—that is to say, the significance that is attributed to the first arrow. The graphic presentation of Bernard's schema could make us believe that H follows O by a constraining local process, especially by the rules of induction. Bernard, however, rightly criticizes the legitimacy of inference by Baconian induction and argues that hypothesis is, certainly, based on a knowledge of the "facts" but does not necessarily result from it. Hypothesis is not inferred from the facts but, according to Bernard, produced by the experimental researcher's "sentiment" (that is to say, by intuition) by a constructive mental activity that depends both on known facts and on an intuitive forecast of facts to be known. The researcher is—says Bernard—nature's "foreman."[35]

Similar opinions have been expressed by the chemist Justus von Liebig (1803–1873) and by the philosophers William S. Jevons (1835–1882)[36] and Émile Meyerson (1859–1933).[37] I shall return to them concerning various myths about the origin of scientific hypothesis. For the moment, what concerns me foremost is to demonstrate the illusory character of the usual interpretation of Bernard's schema that makes observation and theorizing alternate in a clear-cut and absolute way.

In a study of the origins of the notion of living fiber and of the cellular theory, I paraphrased Virchow's dictum by stating: "*omnis theoria ex theoria.*"[38] In my opinion, there is no simple induction from facts to theories and there is, moreover, no creation of hypotheses ex nihilo. One always passes not from observation to theory but from one theory to another. This passage, however, takes place under the influence of observations or through indirectly becoming aware of new "facts." Contrary to Popper, I believe that this influence is not only a falsifying one—that is to say, an eliminator of hypotheses at variance with reality—but also in some way a positive theorization leading toward ideas that have a relatively high chance of survival.

My interpretation of the concatenation characteristic of experimental reasoning differs from Claude Bernard's schema: It is a sequence of theoretical views in a state of flux that produces experiments and is modified following their results. Figure 1 is a symbolic representation of it.

$$E_0 \qquad E_1 \qquad E_2$$
$$\downarrow \qquad \uparrow\downarrow \quad \uparrow\downarrow$$
$$\rightarrow H_0 \rightarrow H_1 \rightarrow H_2 \rightarrow$$

FIGURE 1. H_0 = initial theoretical framework; E_0 = initial observation, not necessarily caused by theoretical considerations but acting only through its integration with H_0; H_1 = initial theory modified by a new hypothesis; E_1 = experiment initiated to ascertain the accuracy of the logical consequences of H_1

10. The Myth of Baconian Induction

It is rare today to find philosophers of science who wish to rank induction as the principal procedure in the logic of discovery. Although there are still scientists and logicians who endeavor to save this myth, numerous lucid critiques, from Hume's to Popper's, allow me to restrict my own appreciation to a few rather summary considerations. I shall not, therefore, attend to the problem of the logical foundation of induction, or to its probabilistic aspects, or to paradoxes of confirmation by the inductive method.

One first difficulty of complete inference from empirical data is due to the impossibility of obtaining them in a raw state and of systematizing them according to their properties alone. What one notices, through observation (i.e., as a "fact") depends not only on properties of objects (or, in other words, sense data) and on some epistemological[39] categories with a general validity alone— as radical empiricism suggests—but also on the theoretical framework prior to apperception and to the integration of the "fact" in our world vision. Hanson has brilliantly demonstrated that "facts" are in mutual dependence with the formalism that expresses them and are impregnated with theoretical prejudices.[40] A two-way link exists between the results of observation and language. The idea we have of an instrument and the theory that gives birth to an experiment influence the way facts are noticed and described. Looking down an optical tube one sees something else, another "fact," as one takes such an instrument for a microscope, a telescope, or a kaleidoscope.

To illustrate the role of prejudice in biological discoveries that, at first sight, are not far above the level of simply noticing the "facts," let us recall the history of *Bathybius haeckelii*, of which Rupke[41] has recently published a good analysis. In the attainment and disclosure of that mistaken discovery, in 1868, of an extremely primitive animal (it was actually an inorganic formation), psychological and even epistemological factors intervened that biased the description and interpretation of relatively simple phenomena. Those factors notably included the enthusiastic acceptance of a certain form of Darwin's theory and the need to bring new evidence into a scientific debate with a strong ideological resonance.

Mendel's discovery is sometimes quoted as a typical example of inductive empiricism. Yet, thoroughgoing analyses of Mendel's work carry the conviction that the logical unfolding of his demonstration should not be conflated with the actual process of his discovery. The fundamental hypothesis on the segregation and statistical distribution of characteristics has not been drawn, by simple inductive inference, from experimental results. According to the biologist and statistician Ronald A. Fisher (1890–1962), the results obtained by Mendel are "too good" (a proportion of 3.01:1 noted on 8000 peas)—that is to say, rather improbable without the regulating interference of an underlying hypothesis, of an anticipatory knowledge of the 3:1 ratio.[42] That is not all:

Mendel's original results in experiments on the hybridization of descendants with parents are both too good from the point of view of his own final theory and false according to our present knowledge (it would have been necessary to obtain 37 percent, instead of 33 percent, of recessive-character representatives). Therefore, Mendel was expecting certain numerical data; he was testing his hypothesis and did not infer it from observed facts. However, certain new historical and mathematical precisions, due especially to van der Waerden and Weiling,[43] seem to temper Fisher's conclusion according to which Mendel first formulated his theory and then performed his experiments. Theorizing and experimenting went hand in hand: The essence of Mendel's discovery was acquired during the carrying out of the experiments.

Granted, there are no logical rules to go with certainty from the particular to the general, no impeachable logical foundations for inductive reasoning. But must one go as far as Popper and say that induction is an illusion in the strongest sense of this term—that is to say, a mental process without logical foundation or, as it were, without existence? According to Popper, induction exists only in appearance, thanks to the selection of hypotheses obtained by a completely different process. An assertion aspiring to general validity would then be in no way influenced, in its genesis, by observed facts. It would secondarily be brought in harmony with these and not be inferred from them.

For a historian of science, convinced as he is that reality ranges wider than the excessive rigidity and extreme simplifications of logical patterns, it appears rash to reject induction in so drastic a fashion. While admitting the nonvalidity, in pure logic, of inference from the particular to the general, we are not forced to deny the existence of a mental inductive process that generates hypotheses without, however, justifying these. If there is no induction in the logical sense of the word, it may well be that there is a kind of psychological inference, an orientation of thought via knowledge of empirical data.

The "discovery" of the "cell" by Robert Hooke (1635–1703) and the progress of micrographical studies in the seventeenth and eighteenth centuries did not lead directly to the formulation of the cell theory. This is a historical example of the impotence of the inductive method, such as it was taught, precisely at the time in question, by the brilliant English chancellor. Various theories on fibers, globules, and "molecules," understood as elementary carriers of life, did not give birth to the authentic cell theory before Matthias Schleiden (1804–1881) conveyed to Theodor Schwann (1810–1882) certain data gathered from observation. I interpret this historic case as an instance of subtle dialectic between attempts at generalizing theorization and concrete observation.[44]

11. The Illusion of the Fundamental Epistemological Role of Verification

We catch a glimpse of this old illusion in the common parlance of scientists who, even today, speak more often of verifying than of falsifying hypothetical propositions or descriptions of the observable. The tendency to the preservation of the structures of the acquired knowledge, to the minimal alteration of these in the process of integrating new knowledge therein, appears as a fundamental characteristic of our mental functioning. This is an "epistemological obstacle," in Gaston Bachelard's meaning of the term.[45] One's primary desire is to see knowledge increased, not overturned. Having stressed that the underlying theoretical framework influences observation and determines certain aspects of the "facts," I should have specified that one is obliged to see *in relation to* what one knows and one tends to see *in keeping with* what one knows.

In my study of Claude Bernard's toxicological experiments, I pointed out a particular kind of epistemological obstacle that, for a researcher with original and profound views, makes it very difficult, in a later moment, to falsify and forsake what was, earlier, the essence of his own discoveries. There is, without doubt, a preference of a psychological nature for verification. However, we must not conflate psychology with logic: Preference at the one level in no way means superiority at the other.

The test to which a scientific assertion is subjected, either by examining its coherence within a system or by confronting its deductions with observable data, must not prejudge the positive or negative character of the result. This seems to indicate that, from the point of view of method, there is no difference between verification and falsification. Indeed, a good many researchers have believed, and some still do, in the myth of epistemological symmetry between verification and falsification. Such a myth does not stand up to logical analysis, not even to a few rather superficial considerations. For assertions with a general validity, falsification is obviously stronger by its consequences than verification: The first establishes nontruth, while the second does not establish truth. Falsification appears as definitive; verification is only provisional. Certain logicians prefer, therefore, to speak of confirmation, a concept that allows a probabilistic relativization of verification.

In short, falsification of a scientific theory seems to be a condemnation without appeal, while verification would be but a precarious acquittal. By the logic of falsification, one aims at a definitive epistemological judgment; by the logic of confirmation, one tries to secure a process of progressive approximation to truth. Karl Popper, who has spotlighted the epistemological superiority of falsification, upholds the thesis of the completely illusory nature of verification (even in its refined form of relativized confirmation), which is, according to him, nothing but a series of aborted attempts at falsification.

12. The Illusion of a Perfect Asymmetry between Falsification and Verification

Faced with the logicians' method of reasoning, one is forced to admit that scientific theories are not verifiable in the absolute sense of the word. Historical examples of theories that have never been falsified evidently prove nothing: We do not know their future. But a different demonstrative importance would attach to the historical existence of theories that, after being falsified, have later been recognized as "true." Such cases are known; indeed, they are quite numerous. So, to my surprise, historical inquiry seems to show that, in fact, scientific theories are not falsifiable either in a definitive way.

The logical explanation for this can be easily found: Data provided by observation can be improperly noted because of technical or psychological errors and, more important, the reasoning process that leads to falsification always presupposes tacit acceptance of a conceptual framework that may be called into question after the event. Such an alteration of certain tacitly accepted presuppositions leads to a disavowal of falsification: Which explains, on the one hand, the historical possibility of a rebirth of scientific theories judged at a given moment as certainly false and, on the other, the impossibility of *experimentum crucis*.

Let me quickly look at a historical example: choosing once again, to make our analysis easier, the discovery of the circulation of the blood. Harvey was at grips with serious difficulties and could not decide, for many long years, to publish his new physiological system. Without a knowledge of the blood capillaries and pulmonary alveoli, it was impossible to explain properly the passage of blood at tissue-level (from the arteries to the veins) and at lung-level (from the veins to the arteries). Assuming the existence of blood capillaries, invisible to the naked eye, Harvey acknowledged implicitly the Galenists' right to believe in the existence of invisible pores in the interventricular partition of the heart. In other words, he weakened an important criticism of the ancient system. To explain the venous return to the heart, one had to turn to nonmechanical conceptions (the spontaneous return of blood to its natural place) borrowed from an outdated (*périmée*) physiology.[46] The origin of the motor power of the heart remained an obscure problem. But what must be remembered above all in the history of Harvey's theory is that, in the context of the positive knowledge of his time, it was "falsified" by the attacks of Johann Vesling (1598–1649) and Jean Riolan the Younger (1577–1657). For his part, the English physiologist disproved the Galenic theory (especially by his quantitative considerations); all the same, he was unable to reply to the serious objections of his opponents. On a practical level, clinical failures of blood transfusion—that is, of the first therapeutic method based on the new theory—weighed heavy. In the purely physiological field, a major difficulty lay in the following problem: how to explain the difference between arterial blood and venous blood if it is the same blood that circulates continuously through all the vessels. Replying to this, Harvey had to maintain that such a difference does not exist in the body and that it results from particular conditions in the observation of blood issuing from injured vessels.

We know today that these refutations of Harvey's theory are illusory, but we know too that that could not be demonstrated at the time and, thus, that Harvey had to turn to very weak ad hoc hypotheses. In the eyes of many seventeenth-century doctors, his theory had been falsified; its success among the savvy ones was only possible thanks to the undeniable falsification of the opposite theory.

Of course, such historical situations are not unknown to present-day upholders of falsificationism (see Antiseri's lucid and instructive analysis).[47] "Ingenuous falsificationism" is the term applied now to the belief in the possibility of definitive refutation of a theory with empirical content. Popper and his followers take a subtler position, an attitude of "methodological falsificationism," that recognizes the impossibility of condemnation without appeal ("disproof") and skillfully uses relativized concepts, such as "illusory refutation" and "ad hoc hypothesis."[48] Each theory can be "immunized" against formal disproof by auxiliary hypotheses, but, curiously enough, as it becomes more and more unassailable, it does not improve its chances of being true. A scientific theory that accumulates saving hypotheses and presents itself in a more and more impregnable form becomes more and more suspect and may even pass out of the field of science altogether. As there is no sound method of distinguishing, in the heat of a scientific debate and without historical perspective, an "ad hoc hypothesis" from a "justified complementary explanation," the asymmetry between verification and falsification is, in practice, less important than it seems at first sight.

The historian who studies the process of scientific discovery is often struck by the precarious character of the first "good" hypotheses in the solution of a problem. They are often "falsified" out of hand, yet their discoverers do

not abandon them but cling to them and end up falsifying the first-stage falsifications. The best hypotheses often contradict one part of the experiment:
They call for a re-examination of the "facts" that, at that point in history,
seemed established in definitive fashion. The strength of a new theory does
not, therefore, lie solely in its resistance to attempts at refutation. There must
be, particularly in the initial phase of a discovery, factors that incline one in its
favor and that are irreducible to the single-handed logic of falsification.

Why out of two (or several) nonfalsified theories, or two (or several) theories historically falsified and saved by auxiliary hypotheses, does one decide in
favor of one rather than the other? Neopositivists put forward the criterion
of probability, especially the degree of corroboration.[49] Popper appeals to the
principle of maximization of empiric content.[50] Serious authors have spoken
of factors of an aesthetic nature, of "poetical," even "mythical," preferences.
I shall draw attention to a characteristic that could be important for the decision scholars make (without believing, however, that it is the only one to turn
the scale): the "perspectival" character of a theory. For a historian of science,
to abandon a scientific theory and accept another is more often a matter of
methodological opportunity than of strict refutation. Formal logic intervenes
generally after the event. It justifies more than it decides the way of scientific
research.

13. The Myth of the Strictly Logical Nature of Scientific Reasoning

Now, as a result of my last statement, we find ourselves faced with the particularly delicate problem of the relationship between the logic, psychology,
and sociology of scientific discovery. I do not have time to analyze properly,
in this context, the three myths that consist in reducing the total reality of
"science in the making," or at least its essence, into just one of these three
approaches. Let us be satisfied with a quick overall view, hoping to be able to
deal with this subject more fully in the course of another seminar.

My account thus concludes with a summary and short denunciation of
three illusions that, because of their methodological importance, should have
been placed at the head of a list of errors to be avoided and that, by their
special epistemological nature, would have deserved the detail and finesse of
a well-documented monograph.[51] The three myths in questions have a common source: the dissociation of the unitary triad made up of the three aspects
of acquisition of knowledge, corresponding to the constitutive triad of the
concept of man (see Figure 2).

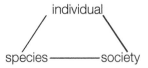

FIGURE 2

(Concerning the latter, see Edgar Morin's publications.)[52] The distinction between the logical, the psychological, and the social lies not in reality but in our means of intellectual analysis. It is a necessary and fruitful process of dismantling, but it creates difficulties and exposes one to the risk of mythicizing, in an exclusive, imperialistic fashion, each of the components of the global system. The three approaches are concurrent, mutually antagonistic, and complementary. They should not be impervious to intercommunication.

In the analysis of scientific discovery, two key concepts are used: *induction* and *intuition*. These are, in my opinion, "amphibological" concepts that, because of their polyvalence, have a significant operational value and, at the same time, are the despair of certain philosophers or psychologists who would like to lock them in the field of a single investigatory discipline. Induction is a concept invented by logicians, and yet they have never succeeded in reducing it, in a completely satisfying manner, to formal logic. Factors of a psychological nature intervene in inductive inferences. As for intuition, while it is a concept belonging to the field of psychology, it cannot be successfully studied if one does not take into consideration the categories and rules of logic.

Positivists with varying allegiances and neo-Baconians still fight bitterly in favor of the old rationalist myth that reduces scientific investigation to a practical application of logic. It is an undeniable fact, which we must not forget in this debate, that a considerable number of scientific discoveries result simply from the application of an impeccable logical reasoning to concrete problems. In the genesis of such discoveries the historian can easily grasp the "guiding idea" and is bound to admire the precision of observation (Leeuwenhoek, Redi, Faber, etc.) or the persistence in the application of certain clear, relatively simple hypotheses (Pasteur, Ehrlich, Ludwig, etc.).

When it is a matter of "deductions"—that is to say, applications of a theory or an investigatory technique to cases not previously contemplated—rationalistic explanation is triumphant. Ascertaining the origin of a completely new idea is a more complex matter. It seems that to explain the genesis of certain discoveries, and precisely the most original and fruitful ones, it is necessary to refer to irrational processes. It goes without saying that the term *irrational* does not indicate anything "supernatural," magical, or unaccountable in its principle. I only wish to convey by this word that a particular creative element intervenes in the roots of rationality itself and that scientific thought transcends the limitations of classical logic.

The strongest condemnation of the myth of the sovereignty of logic has been uttered by none other than certain logicians themselves, anxious to circumscribe their territory and abandon litigious fields to feel more comfortable in their fortress. Does not this withdrawal go too far? It seems dangerous to me to affirm that intuition (that is, the actual genesis of discovery) eludes a really scientific study for the reason that it is outside the framework of logic in the narrow sense of the word (that is, of present-day formalized logic). Popper considers that "there is no such thing as a logical method of having new ideas or of a logical reconstruction of this process" and that "every discovery contains 'an irrational element' or 'a creative intuition,' in Bergson's sense."[53] The act of conceiving or inventing a scientific theory, just like the birth of a musical theme, cannot, by this reasoning, be the object of logical analysis but comes within the purview of empirical psychology.

This opinion seems to me, on the one hand, well entrenched against any formal attack and, on the other, very dangerous inasmuch as it seals off certain paths of inquiry. To criticize it, one would have to redefine the conceptual framework of classical logic. The logic of scientific discovery (if there is one, as I believe there is) cannot be the sort of formal logic that is capable of dealing exclusively with matters of justification or validation. In this sense, the title of Popper's main work is a false promise.

14. The Myth of the Strictly Irrational Nature of the Origin of Discoveries

This myth rests, on the one hand, on a strict and rigid rationalism—that is to say, on the considerations of the logicians I have just mentioned—and, on the other, on a romantic, irrational, sometimes even surrealistic idealism. These last tendencies are in fashion again today.

A psychology of scientific discovery, which invokes in a quasi-mystical manner the "personal genius" of the researcher, must be evaluated with the greatest severity. Even in its variants bearing semblances of scientificity (the analogy, for instance, between the birth of a new idea and biological mutation [Charles Nicolle] or the quantum leap, speculations of certain psychoanalysts and biographical studies with a psychological veneer), such an approach suffers from a fundamental flaw: that of untestability (*incontrolabilité*). "Socrates's demon" is an amusing hypostasis of the actual experience of inspiration, of sudden illumination, but is also the prototype of "mythological" explanations. One must, in this day and age, reject as illusory any explanation presupposing the existence of "supermen," of exceptional geniuses thinking and acting in accordance with modalities differing qualitatively (and not just quantitatively) from those of the common run of mortals.

Empirical psychology can certainly shed a precious light, which is indispensable for understanding creative thought. Let me recall the results of two classical methods: that of introspection (see J. Hadamard's excellent work on the subject)[54] and that of the observation of a subject under experimental conditions (from the already old and yet still relevant works by Claparède,[55] by Duncker,[56] by Maier, to more recent research by Wertheimer,[57] Bruner,[58] Gruber,[59] and so many others). The "problem-solving" type of experimental analysis has proved successful with regard to the elucidation of certain elementary processes of reasoning, but we are still far from being able to draw from it definite, useful conclusions concerning artistic invention or scientific discovery. In the laboratory, creativity is limited in time and applied to simple problems, which does not therefore imitate the actual conditions of scientific research. A psychologist, who is concerned with scientific creativity at the highest level, must turn toward historical documentation, for at the present time obstacles of various kinds prevent psychological experimentation on renowned scholars grappling with important scientific problems.

15. The Myth of the Sociological Explanation of Scientific Discoveries

Far be it from me to wish to deny the impact of socioeconomic factors on the progress and vicissitudes of scientific thought, as much at the individual level of each scholar as at that of the community giving rise to, and encouraging, the solutions that we feel are needed; it appears nevertheless illusory to seek, at all costs, in these factors the complete explanation of creative scientific activity and its results. In the history of science, "externalism" is a mirage and, pushed to its limits, an untruth.

There is probably a connection between Harvey's monarchism and his ideas on the central role of the sun and the heart, between Magendie's revolutionary education and his iconoclastic pragmatism, and between Virchow's middle-class liberalism and his cell theory; but the presence of ideological elements in a scientific work, and even in any scientific work, is something different from the reduction of science to ideology.[60] If it is true that scientific opinions are most often in agreement with the political stands of scholars, it does not follow that it is possible to deduce such opinions from ideology, especially when we deal with original ideas. Still more illusory have been attempts to explain the genesis of scientific theories through the direct influence of socioeconomic factors. The achievements of men like Mikhail Lomonosov (1711–1765) or Nikolay Lobachevsky (1792–1856) are, certainly, influenced by the economy of Tsarist Russia but are not epiphenomena of it. It is, without doubt, very illuminating to detect and follow up as far as possible the connections between the material determinants of general historical development and the meanderings of scientific thought, but provided one does not willfully close one's eyes by halting inquiries of a nonsociological nature.

The difficulty and the appeal of the problem of scientific discovery are derived to a large extent from the fact that it is placed at the crossroads of several disciplines—in at least two ways. First, in spite of the diversification of present-day branches of sciences and techniques, the fundamental problem of the creative activity of the human intellect is everywhere the same: The act of achievement of a scientific theory or technical invention is analogous, if not identical, to the act by which a work of art is created. Secondly, the position of this problem in the theory of knowledge seems to be so central that it constitutes the Gordian knot of all possible approaches, and no unidisciplinary method of inquiry can give, single-handedly, results that are entirely satisfying.[61]

PART FOUR

Memoricide: War and the Eradication of Cultural Memory

EIGHT

A Memoricide

> For historian Mirko Grmek, the bombing of the historical site of
> Dubrovnik . . . was not an isolated act, but was part of a broader political
> strategy aimed at systematically destroying all traces of the Croatian past.
> Grmek called this politics "memoricide,"—a term he coined to echo that
> of "genocide."
>
> —LOUISE L. LAMBRICHS, "PRESENTATION
> OF 'UN MÉMORICIDE'"

On the pretext of a hypothetical genocide Croatians ostensibly intended to
perpetrate on Serbians, the so-called Federal Army and the leaders of the
Republic of Serbia have initiated—for real—the genocide of the Croatian
people. The rumors that they sparked were followed by horror, massacres,
and mass destruction of Croatian cities and towns. Regions where the Serbian minority lives are relatively poor; they are only interesting to Belgrade
politicians insofar as they provide legitimacy for occupying a space considered "vital" for Serbian people's expansion and flourishing. In fact, they want
to conquer rich territories such as the Dalmatian coast and the occidental
Slavonic plain, which are ethnically and historically Croatian. To this end,
Serbians have to make tabula rasa in the military conquered territories, scare
away its population (to prevent its return), and eliminate material traces of
Croatian nature from all these regions.

Between July and November of this year [1991], more than four hundred
monuments or sites of historical and cultural significance have been destroyed
or badly damaged, including many Catholic churches, tens of monaster-
ies, museums, libraries, archives, archaeological sites, as well as historically

important castles and lordly houses. Some of these churches were of major artistic value as World Heritage sites; others, which saw the passing of historical
events and were privileged places of cult, were particularly dear to Croatians;
all of them could bear witness to the presence of Catholicism that the Serbian
league of communism and orthodoxy would like to "uproot."

Pillage of the Heritage

Everything happens as if the Serbian army generals chose their targets using a
trustworthy cultural history textbook of Croatia. The so-called Federal Army
has indeed bombed artistic masterworks such as the cathedral of Šibenik and
Osijek, the churches of Saint Anastasia and Saint Chrysogonus in Zadar, in
addition to Saint Blaise's place in Dubrovnik. It also fired at Nin, a small vil-
lage, petty from a military point of view, but which represents the cradle of
medieval kingdoms in Croatia. In bombing Split, armies targeted with great
precision the museum of medieval Croatian art.[1] The library and national
archives of Slavonia have also been destroyed, cemeteries were shelled, and
funeral monuments were demolished. The Trsteno Arboretum, which is a
valuable reservation of rare species, was burned down on purpose, and the admirable park of Lokrum is presently on fire. The lazaretto of the world's most
ancient quarantine in Dubrovnik was badly damaged, just like the Domus
Christi, one of the first modern hospitals in Europe. A great many of those
acts of hooliganism, I should emphasize, were perpetrated after the occupa-
tion, not during fights.

The pillage of heritage sites is now escalating in a strategy of total war, and
commentators speak of it in terms of a "cultural genocide." It seems more
useful to name this new concept using a new term: In this respect, I suggest
that *memoricide* is a concept more appropriate to describe this reality, knowing
that in ancient Latin, the term *memoriae* means not only memories but also
historical monuments.

Dubrovnik: The Slavic Athens

Some historians of science, more interested in the social conditions of scientific discoveries than in the logic of research, separate producers and users of scientific and technical knowledge.[1] Although attractive, this dichotomy should not eclipse the important role of cultural centers in promoting the transmission of knowledge.

Operating as a critical junction between Western and Eastern Mediterranean cultures over the course of several centuries, the history of Dubrovnik, the small republic of Ragusa, is a paradigmatic example of such a cultural center. On their way to the hinterland of the Balkans and the Middle East, scholars, medicines, and scientific instruments coming from Italy, as well as Western books, had to travel through Dubrovnik. Similarly, the city long served the migration of Oriental scholars and ideas on their way to Western Europe. From the Middle Ages until the present, however, this intellectual traffic coming from the East progressively declined. Merely a bridge between two civilizations, the Croatian city of Dubrovnik was effectively turned into a beacon of Western thought.

The Historical Context

Built on an exposed rock with no land to plough within easy reach, life in Dubrovnik was almost entirely turned toward the sea, as it was in Venice or Genoa. Fortunately, the city was also flanked by thick forest, which could be transformed into ships, and stood at the borders of at least two civilizations. Connecting Latin and Byzantine parts of the ancient Roman Empire during the twelfth century, Catholic and orthodox religious worlds later on, and Western European civilization and the Turkish empire still more recently, the geographical position of Dubrovnik was long the ideal interface between West and East.

An ancient autonomous town, Dubrovnik maintained its privileges when brought under the influence of Byzantium, as well as when it was dominated by Venice in 1205. In 1358, when it became rightfully autonomous, the city acquired the legal status of a suzerain republic. While recognizing officially the supreme authority of Hungarian and Croatian kings, as well as that of the Great Door, the Republic of Saint Blaise continued paying steady tributes to various sovereigns and conducted its own politics, which were entirely driven by the social and economic interests of the local oligarchy. The dislocation of this enclave of freedom in 1808 (which was later integrated with the Hapsburg Empire) in the name of liberty was neither Napoleon's greatest feat of arms nor indeed the most successful political action of French foreign policy.

Free medieval states in this part of Europe went out of existence following the Ottoman conquest of territories known today as Serbia, Macedonia, and Bosnia and Herzegovina, and the political dominion Austrian and Venetian Republics exerted upon Croatia and Slovenia. Until the nineteenth century, Dubrovnik remained, in fact, the only political entity among Southern Slavic countries able to preserve its autonomy and liberty, which, as expressed by Ivan Gundulić (1589–1638), Ragusa's most famous poet, is "the treasured gift that God to us endowed."[2]

Once governed by some tens of families who followed very strict rules of conduct, and in contrast with other great feudal lords in the country, Dubrovnik's patricians did not primarily derive their economic power from the land. Leading sources of income included trade, bank transactions, ship building, and the development of craftsmanship. Not acquired by force, and not even retained by arms, these people's liberty was largely because of their ability to make the best of various conflicts in which powerful neighboring states were opposed to one another.

The Testimony of Boscovich

Like Geneva later on, Dubrovnik was the cradle of many scholars and writers recruited into the ruling class. All of them remained very much attached to their city; and even those whose careers were conducted abroad proudly claimed their Ragusian origin. Let us consider the cases of the physicist Boscovich and the physician Baglivi.

Roger Joseph Boscovich (1711–1787) was a physicist, astronomer, and philosopher; he was also the brilliant author of a wholly dynamic atomistic theory.[3] In *Philosophiae naturalis theoria redacta ad unicam legem virium in natura existenium* [*A Theory of Natural Philosophy*] (1758), he postulated the existence of nondimensional atoms in a relative space (i.e., a force field).[4] Boscovich's curve, defining the relations between two points in such a field, was intended to account for laws of gravitation and mechanics as well as chemical affinities. Boscovich's work dethroned both Descartes's concept of extended matter and Newton's concept of mass. Calling into question the notion of absolute time thanks to an extraordinary intuition, he explained forces of inertia by appealing to the influence of stellar matter on space properties.

Educated with Jesuits in Dubrovnik and later in Rome, Boscovich was a European scientist. Fluent in many languages, he was as much at ease in Milan or Vienna as in Paris or London. A member of the Academy of Sciences in Paris, he died a French citizen. It is a peculiar sign of the cultural politics pursued since the beginning of this century that French dictionaries introduce him as a "Serbian" scientist and philosopher, or sometimes as a "Yugoslav," without even realizing the difficulty conveyed by the term "Serbian Jesuit." Boscovich never forgot his city of birth and long corresponded with his sister, the Ragusa poet Anica Boscovich (1714–1804), in a language that she named "Croatian." Until the end of his life, Boscovich faithfully served the small republic as its unofficial representative in diplomatic affairs. In a commentary of his sister's poem on solar and lunar eclipses, Boscovich wrote (in French):

> On this occasion, I briefly but expressively make a eulogy of my country, bequeathing it with the greatest and truest praises, as those who read the Geographers' works and the literary monuments of Ragusian people know. Surrounded from all sides by barbary and grossest ignorance, we cultivate the exact sciences and especially the arts [*belles-lettres*], and we do so as much in Latin as in Illyricum, which is the language of this country, and with all eagerness possible. Without restricting myself to ancient writers such as Marin Ghetaldi [1568–1626], a famous geometer at a time few people studied such an honorable science; or

Etienne Gradi [1613–1683], an elegant Latin poet, even in the past century when good taste was in decline and arts were degenerating in Italy; or even P. Anselmo Banduri [1671–1743], a known Benedictine monk because of his many books printed in France, and without even mentioning great many others who flocked to me. Among living authors, I shall only name Benoît Stay [1717–1801], who stands for a thousand others, in all kinds of knowledge or in Latin poetry. . . . As to national language, we have two excellent epic poems, *Hosmaniade* [from Gundulic´] and *Christiade* [from Maurice Vetranovic, 1482–1576], as well as a great many of other poetic works that are most valuable. Among these, the poetries of P. [Ignace] Giorgi [1675–1737], a Benedictine abbot who earned a great reputation in this century thanks to other printed Latin books, filled with erudition, stand out singularly.[5]

Let me also mention Benoît Cotrugli (1400–1468), another Ragusa scholar, to whom we owe a book on the art of trade written in Dubrovnik in 1458 but published only in 1573 in Venice, which is the first systematic treaty on principles of accountancy. The writings of Félix Petancius (1455–1518), a Ragusa humanist, are also quite significant. His book *Historia Turcica et Descriptio Turcia* is the first Western study on this emerging civilization.

The Medical Vocation of Dubrovnik

Giorgio Baglivi (1668–1707) also exemplifies the patriotic affection of Dubrovnik's scholars.[6] In addition to being a professor in Rome, Baglivi reformed clinical medicine, was a leader of iatrophysicians, and promoted physiology and pathology as sciences based on properties of living fibers understood as elementary structures. The renown practitioner was physician of the pope and a skilled experimenter who wrote in a laconic style (according to historian of medicine Ch. Daremberg, Baglivi was "the most sensible and the Ciceronian of all iatromechanicians")[7] and he did not hesitate to praise his homeland: "Ragusa, charming and very noble city of Dalmatia once named Epidaurus and famous for its temple of Aesculapius, today the capital of a free and eminent Republic. I was born in this city on September 8th 1668 at the sunrise."[8]

This text deliberately mingles the names of Epidauros in Greece with Epidaurum in Dalmatia (today Cavtat), a colony of the native founders of the medieval city of Dubrovnik. Baglivi's attachment to classical tradition is what seems most significant to me here. In the sixteenth century, Ragusan peoples sculpted the figure of Aesculapius on one of the columns of the government palace, depicting him as a bearded scholar surrounded by alchemical instruments. Legend has it that a temple dedicated to this divinity once stood there.

Be that as it may, the city always wanted to remain faithful to its medical tradition. According to its pragmatic administrators, it was clear that the State was responsible for the care of the health of its citizens in the same way it behooved it to maintain public order, to manage international relations, and to attend to commercial interests. Since the fourteenth century, Dubrovnik had an admirably well-organized public health service. In 1377, a decree of the Great Council created the first quarantine in the world conceived as a protective measure against plague. Before this date, some other towns resisted the disease by isolating sick individuals and by interrupting trade with territories affected by the epidemic. Ragusan patricians, however, could not bear ending commercial relations during epidemics; they thus settled for temporarily isolating individuals who were not yet ill but who were suspected of incubating the disease.[9]

The republic of Ragusa never intended to create a university on its territory. For reasons of internal politics, it kept local doctors, solicitors, and schoolteachers at its service, many of whom were educated abroad and came from abroad. A number of these scholars, in fact, were Marrano looking for shelter. Their presence contributed to the mixing of ideas and it did so without them being perceived as threats to the privileges of the ruling class. Physicians in the service of Dubrovnik often went into Turkish territories as pashas and even the court of Istanbul frequently called upon them for consultation. It was thanks to one such practitioner that in 1467 the sultan received several Latin medical manuscripts as gifts from the Dubrovnik Senate.

Ragusan trade with its neighbors was flourishing. The principal exports were ore, skins, wood, and other raw materials, while imported products included not only fabrics, oil, salt, and arms but also books, drugs, and, indeed, the most diverse tools. Ever since the Middle Ages, the Republic had a communal drugstore that was especially well provided. Goldsmiths from Dubrovnik produced and exported jewels, clocks, as well as all sorts of astronomic instruments.

The Historical Sense of the Actual Bombing

Called the "Slavic Athens," Dubrovnik symbolically represented freedom and prosperity, at least in the mind of South Slavic people. Because it was a Catholic and prosperous city, however, it also looked degenerate, enslaved by the forces of evil, and even corrupted by false knowledge to more orthodox peasants.

When Serbian political and military leaders decided to bomb the city in October 1991, no barracks lay behind its ramparts, no heavy weapon was hidden there; and yet, these lords of war were right on target. Bombs fell on the sumptuous Catholic churches and on the vestiges of an ancient synagogue. Touching one of the oldest libraries of southeast Europe, the bombs also damaged an ancient aqueduct, a marvel of Renaissance technology, and brushed against buildings that once housed the world's first quarantine lazaretto. Likewise, shells hit one of the first occidental hospitals, which, in contrast to other similar European institutions, already provided permanent medical service in the Middle Ages and bestowed as much importance on keeping the body in a state of health as on the salvation of the soul. The library and study rooms of the postdoctoral Inter-University Center were burned down and completely destroyed. Was it a matter of chance that such a fate happened to the seat of an institution worldly known as a citadel of pacifist movement and instigator of collaborations between universities of various countries?

Serbian and Montenegrin soldiers came down from the mountains stealing televisions, household items, and other goods of civilization they piled in houses outside the city's walls. But in perpetrating such a barbarian incursion their leaders' goal was different: They were deliberately targeting the utmost symbol of Western tradition for their people. They wanted to demonstrate with a concrete example that the Western world did not, or could not, defend its traditional values with anything more than soothing words and inconsequential threats.

What Future for Dubrovnik?

The Serbian leaders who extol politics of memoricide[10] are not misled when they continue, as of today (April 1995), to bomb, from time to time and in a highly symbolic way, this "open city" classified as a World Heritage site by UNESCO. Today, Dubrovnik houses archives that are of central importance for the history of the Mediterranean Sea, several scientific institutes related to the Croatian Academy of Sciences and Arts (the Institute of History, the Institute of Biology, and the Institute for the Study of Marine Corrosion), and an inter-university center that regularly organizes symposiums and advanced summer courses. Destroyed by the Serbian bombing, this center is now in the process of being restored and could become once again a welcoming hive for researchers from all over the world. We should take advantage of the actual rebuilding of the city to create an international artistic center. Dubrovnik would thus recover, though in a new way, its old vocation of being an interface between civilizations.

ACKNOWLEDGMENTS

This project emerged while I was flying from Geneva to Montreal in May 2012 to attend a conference on the philosophy of science. Remembering a conversation with Stefanos Geroulanos and Todd Meyers earlier that year in Skagen, Denmark, during a workshop on Georges Canguilhem, I set out to inquire about a collection of essays by Mirko D. Grmek for the Forms of Living series during a stop in London. After landing in Montreal later that day, I was very pleased to find Stefanos's positive response in my mailbox. Since the inception of this project, Stefanos's and Todd's enthusiasm have helped transform a vague idea into a published book. I am extremely grateful to them for their support, advice, and guidance at each step of the project.

The initial phase was carried out when I was a postdoctoral researcher (FRQSC) at the Institute for the History of Medicine and Health in the Faculty of Medicine at the University of Geneva (2012–2013). After joining the Université Laval (Quebec City) in July 2013 as a faculty member, I soon realized that the project would take longer than initially planned. These delays, however, turned out to be for the best. Becoming familiar with Grmek's work and its historical and philosophical background is a lengthy process; acquainting oneself with his extensive correspondence, biographical notes, and unpublished papers deposited in the Fonds Grmek at the Institut Mémoires de l'édition contemporaine, in Caen, France, takes even longer.

I cannot hope to thank everyone who helped me along the way to complete this project, but I will nevertheless try to highlight a few debts I have incurred over the years. In Geneva, my thanks go to Bernardino Fantini for welcoming me at the Institute for the History of Medicine and Health and for sharing his memories of Grmek. For facilitating access to the Grmek Papers in Caen, I would like to thank the following staff members of the Institut Mémoires de l'édition contemporaine (IMEC) for their support: André Derval, Yves

Chevrefils Desbiolles, Claire Paulhan, Marjorie Delabarre, Elisa Martos, and Mélina Reynaud. Of course, without the permission of Louise L. Lambrichs, none of this research would have been possible, so I would like to thank her very warmly for her trust.

The bulk of the project was completed between 2014 and 2017, during four consecutive summer research stays at the Institut des humanités en médecine (IHM) in Lausanne (formerly known as the Institut universitaire d'histoire de la médecine et de la santé publique). I would like to thank sincerely Vincent Barras, the director of the institute, for providing me with such a congenial and productive environment for work. I extend my thanks to the wider research community of the IHM for their input, as well as to the administration and library staff. Special thanks go to Maïka Garcia Casse, for her helpfulness in locating many of Grmek's articles, and Jelena Martinović, for her French translation of Grmek's writings in Croatian.

At the Université Laval, I have benefitted from a number discussions with colleagues and students, especially those who attended the graduate seminars in the history and philosophy of science and medicine. I would like to acknowledge the work of my research assistants throughout this project— Jérôme Brousseau, Kéba Coloma Camara, and François Fournier—as well as the support from the faculty and administrative staff of the Faculté de philosophie. I also thank Anne-France Morand for her assistance in locating the English translations of Galen cited by Grmek.

At Fordham University Press, I am grateful to Stefanos and Todd for welcoming this new book in the Forms of Living series, as well as to Tom Lay, John Garza, Richard Morrison, and Eric Newman for overseeing the early and final editing processes with diligence, care, and professionalism. As this book was accepted for publication shortly before Helen Tartar passed away, I would like to acknowledge her contribution as Editorial Director at Fordham University Press.

In addition to the many friends and colleagues who provided encouragement on this work over the past six years, I wish to thank Jackie Duffin for sharing her personal correspondence with Grmek (her former thesis advisor), for extensive and lively discussions about his work, for helping me with the translation of the first essay included here, and especially for her enthusiasm about this project. I am also thankful to Kathleen Hulley, Donald Landes, and Alex Powell for their generous assistance with the translations and the editing of the Introduction. For reading and providing helpful comments on the Introduction to this volume at various stages of its completion, I am grateful to Vincent Barras, Tatjana Buklijas, Jackie Duffin, Staffan Müller-Wille, Camille Limoges, and Thomas Schlich. In particular, Tatjana's knowledge of former Yugoslavia's political complexities has helped me avoid some historical misconceptions. My understanding of Grmek's life and work also benefited from conservations with François Duchesneau, Jean Gayon, Danièle Guinot, Alexandre Klein, Gérard Lambert, Louise L. Lambrichs, Michel Morange, Stephen Morse, Ante Padjen, Ana Cecilia Rodríguez de Romo, and Jonathan Strauss. I also thank Hans-Jörg Rheinberger for the foreword to the book.

Ideas included in the introductory essay were presented during conferences and research seminars at several venues, including Egenis, the Centre for the Study of Life Sciences at the University of Exeter (May 2014); the workshop "Concepts et méthodes dans l'oeuvre de Mirko Grmek" organized at the Centre interuniversitaire de recherche sur la science et la technologie in Montreal (March 2015); the research seminar "Quelle histoire pour la médecine et la science? Dialogue autour d'une oeuvre" at the IHM in Lausanne (November 2015); the American Association for the History of Medicine annual congress in Minneapolis (April 2016); and the workshop "Retrospective Diagnosis Meets New Technologies" held at the Université Laval in Quebec City (January 2018). For making these different travels and research stays possible, financial support from the Fonds de recherche du Québec Société et culture (FRQSC), the Faculté de philosophie of the Université Laval, the Centre interuniversitaire de recherche sur la science et la technologie, and the Institut Mémoires de l'édition contemporaine is warmly acknowledged.

Finally, I would like to thank my family (Hugues, Marcelle, Mireille, and Catherine) for their steadfast encouragement of my pursuing this project and, above all, Cindy for her patience, love, and support.

FOREWORD

1. [The title of the Sixth Course of the International School of the History of Biology Sciences (June 19–28, 1988) was The Measure of Life: Quantity and Quality in Biological Explanation. It was directed by the historian of science Alistair Crombie (Trinity College, Oxford) although the school program itself was under Grmek's directorship. For details about the program of the Sixth Course, see https://ischiasummerschool.org/past-schools/1988-school.]

2. [The workshop "Théorie et méthode dans les sciences de la vie" was organized by Marino Buscaglia, Bernardino Fantini, and Marc J. Ratcliff. In addition to them, speakers included Vincent Barras, Mirko Grmek, Hans-Jörg Rheinberger, and Jacques Vonèche. A report on this workshop can be found in Marc J. Ratcliff, "Journée d'étude 'Théorie et méthode dans les sciences de la vie,' Geneva, 26 January 1996," Gesnerus 53 (1996).]

3. Hans-Jörg Rheinberger, Toward a History of Epistemic Things: Synthesizing Proteins in the Test Tube (Stanford: Stanford University Press, 1997).

4. Claude Bernard, Cahier de notes (1850–1860). Présenté et commenté par by Mirko D. Grmek, preface by Robert Courrier (Paris: Gallimard, 1965).

5. Mirko D. Grmek, "First Steps in Claude Bernard's Discovery of the Glycogenic Function of the Liver," this volume, Chapter 4.

6. Mirko D. Grmek, Raisonnement expérimental et recherches toxicologiques chez Claude Bernard (Geneva: Droz, 1973); Frederic L. Holmes, Claude Bernard and Animal Chemistry: The Emergence of a Scientist (Cambridge, Mass.: Harvard University Press, 1974); Frederic L. Holmes, Investigative Pathways: Patterns and Stages in the Careers of Experimental Scientists (New Haven, Conn.: Yale University Press, 2004).

7. [On the "practice turn," see Léna Soler, Sjoerd Zwart, Michael Lynch, and Vincent Israel Jost, eds., Science after the Practice Turn in the Philosophy, History, and Social Studies of Science (New York: Routledge, 2014).]

8. Eduard J. Dijksterhuis, "The History of Science forms not only the memory of science, but also its epistemological laboratory," in Critical Problems in the History of Science, ed. Marshall Clagett (Madison: University of Wisconsin Press, 1969), 182.

9. Karl August Möbius, Die Auster und die Austernwirthschaft (Berlin: Wiegandt, Hempel & Parey, 1877).

INTRODUCTION: MIRKO GRMEK'S INVESTIGATIVE PATHWAY

1. Mirko D. Grmek, *Diseases in the Ancient Greek World*, trans. Mireille Muellner and Leonard Muellner (Baltimore, Md.: Johns Hopkins University Press, 1989), 1. Because Grmek grants only a conceptual existence to disease, his position is sometimes referred to as "medical Platonism." See Danielle Gourevitch, "Il est en Grèce une divinité . . . ," *Histoire des sciences médicales* 35 (2001): 347 [my translation].

2. "The term 'illness,'" Grmek writes, "should be used to describe the immediate experience of a sick person, the 'experience' of disease, while the term 'disease' should be reserved for the conceptualization of disease by physicians." Mirko D. Grmek, "The Concept of Disease," in *Western Medical Thought from Antiquity to the Middle Ages*, ed. Mirko D. Grmek, coord. Bernadino Fantini, trans. Antony Shugaar (Cambridge, Mass.: Harvard University Press, 1998), 242. See also Chapter 2, this volume. On the distinction between disease, illness, and sickness, see Jacalyn Duffin, *Lovers and Livers: Disease Concepts in History* (Toronto: Toronto University Press, 2002); Bjorn Høfmann, "On the Triad Disease, Illness and Sickness," *Journal of Medicine and Philosophy* 27, no. 6 (2002). The expression "cultural fabrication" is borrowed from Brent D. Shaw who used it in his review titled "Grmek's Pathological Vision," *Social History of Medicine* 4 (1991): 332.

3. Grmek, *Diseases in the Ancient Greek World*, 1.

4. Grmek's work sits uneasily with Adrian Wilson's distinction between "naturalist-realist" and "historicalist-conceptualist" histories of disease. See Adrian Wilson, "On the History of Disease-Concepts: The Case of Pleurisy," *History of Science* 38 (2000).

5. Mirko D. Grmek, "Le passé n'existe pas," *L'actualité Poitou-Charentes* (1997): 31.

6. "I consider myself to be a citizen of the world and I have no nationalist preferences." Mirko D. Grmek to Jacalyn Duffin, July 29, 1989, GRM2-Da.03, Grmek Papers, IMEC, Caen, France [my translation]. This introduction was written following several research stays conducted between 2012 and 2017 at the Institut Mémoires de l'édition contemporaine (IMEC), at the Abbey d'Ardenne in Caen, France, where Grmek's papers were deposited shortly before his death. All references to this material is given using the number of the box (GRM) and the page number when the reference is to Grmek's biographic writings.

7. Like Grmek, scholars such as Temkin were as versed in the history of ancient medicine as in nineteenth-century physiology. Grmek's originality, however, lies with his broad vision of the history of science and medicine where distinct historical periods serve to enlighten each other, on the one hand, and his close engagement with contemporary and political phenomena such as war and pandemics, on the other.

8. During his career, Grmek published an impressive number of encyclopedia entries. For example, he published 71 entries in Croatian in the first edition of the *Enciklopedija Jugoslavije* (1955–1971); 203 in the second edition of the *Medicinska enciklopedia* (1967–1970); 84 entries in French in the *Encyclopédie in-*

ternationale des sciences et des techniques (1969–1973); and 12 entries in English in the *Dictionary of Scientific Biography* (1970–1978). For a complete list of Grmek's dictionary entries, see Danielle Gourevitch, "Biographie et bibliographie choisie," in *Maladie et maladies. Histoire et conceptualisation (Mélanges en l'honneur de Mirko Grmek)*, ed. Danielle Gourevitch (Paris: Champion, 1991).

9. Grmek's complete biography is available in Louise L. Lambrichs, "Mirko D. Grmek: Bibliographie chronologique 1946–2000," in *La vie, les maladies et l'histoire*, by Mirko D. Grmek, ed. Louise L. Lambrichs (Paris: Seuil, 2001).

10. Mirko D. Grmek, "Prolégomènes à une histoire générale des sciences," *Annales. Économies, Sociétés, Civilisations* 20 (1965): 139 [my translation].

11. Mirko D. Grmek, "Introductory Lecture: What Is History? What Is Science? What Is History of Science?," 1967, GRM 14, p. 8, Grmek Papers, IMEC, Caen, France.

12. Mirko D. Grmek, "Médecine et épistémologie: transformation du savoir sur la santé et la maladie," *History and Philosophy of the Life Sciences* 10 (1987): 5 [my translation]. Grmek made this comment during the opening address of a conference in Perugia (Italy) in April 1985, which could be read as a nod to Michel Foucault (1926–1984), who had recently died. Although Foucault and Grmek both studied under Canguilhem, they almost never cited each other's work. Grmek's comments might also be a reference to the work of Jacques Léonard (1935–1988), a French historian of medicine and advocate of "bio-history." See Jacques Léonard, *La médecine entre les pouvoirs et les savoirs* (Paris: Aubier Montaigne, 1981), 7.

13. Gert H. Brieger, "The Historiography of Medicine," in *Companion Encyclopedia of the History of Medicine*, ed. William F. Bynum and Roy Porter (London: Routledge, 1993), 34. Stephen Boyden, *Western Civilization in Biological Perspective: Patterns in Biohistory* (Oxford: Clarendon Press, 1987).

14. Chapter 1, this volume.

15. Colin Jones, "The Pathocenosis Moment: Mirko Grmek, the *Annales* and the Vagaries of the Longue Durée," *History and Philosophy of the Life Sciences* 27 (2005), 7; William H. McNeill, *Plagues and People* (New York: Doubleday, 1976); Alfred W. Crosby, *The Columbian Exchange: Biological and Cultural Consequences of 1492* (Westport, Conn.: Greenwood, 1972).

16. Mirko D. Grmek, "Discours de Rome," (paper presented at Personnalités, problèmes et méthodes de l'histoire des sciences. Cinquante ans d'une discipline entre science de l'homme et science de la nature, Rome, Istituto della Enciclo-pedia Italiana, June 4–6, 1986), GRM 15, Grmek Papers, IMEC, Caen, France [my translation].

17. Frederic L. Holmes, *Investigative Pathways: Patterns and Stages in the Careers of Experimental Scientists* (New Haven, Conn.: Yale University Press, 2004). Grmek and Holmes met in 1963 at the marine biology station of Banyuls-sur-Mer in southwest France and corresponded during four decades. After this meeting, Grmek introduced Holmes to the unpublished papers and laboratory notes by Claude Bernard, which later served as the foundation of Holmes's own research method in history of science.

18. Though Grmek's work gets mentioned by Brieger, it is not referenced in Allan Brandt's article on "Emerging Themes in the History of Medicine." Also, no entry in *Locating Medical History*, edited by Frank Huisman and John Harley Warner in 2004, concerns Grmek or his contributions, though his name appears in a few essays. For recent works on Grmek, see Joël Coste, Bernardino Fantini, and Louise L. Lambrichs, eds., *Le concept de pathocénose de Mirko Grmek. Une conceptualisation novatrice de l'histoire des maladies* (Geneva: Droz, 2016). See also the contributions collected in the 2001 "Special Issue in Honor of the Late Mirko Grmek" of *History and Philosophy of the Life Sciences*. For a broader assessment and contextualization of his work in medical historiography and philosophy of science, see Pierre-Olivier Méthot, "Mirko Grmek et l'histoire de la médecine et des sciences au XXe siècle," in *Médecine, science, histoire. Le legs de Mirko Grmek*, ed. Pierre-Olivier Méthot (Paris: Matériologiques, forthcoming).

19. Ana Cecilia Rodríguez de Romo, "Eloge: Mirko D. Grmek, 9 January 1924–6 March 2000," *Isis* 92, no. 4 (2001): 742.

20. Grmek was an only child. On his upbringing and early education in Zagreb, see Louise L. Lambrichs, "Un intellectuel européen engagé," in Grmek, *La vie, les maladies et l'histoire.*

21. There is a picture of this self-made laboratory in Grmek's papers deposited at IMEC. GRM 120, album 2 (1940–1944), p. 3, Grmek Papers, IMEC, Caen, France.

22. Grmek's journey to escape from Italy took him through Lucca, Florence, Bologna, and Milan. With two other fellows from the military academy, he crossed over to the Swiss border in November 1943, near Lake Lugano, and arrived at the first refugee camp in Bellinzona.

23. Mirko D. Grmek, *La guerre comme maladie sociale et autres textes politiques* (Paris: Seuil, 2001), 237 [my translation]. Those poems have been translated in French and published in Lambrichs, "Un intellectuel européen engagé," 172–74.

24. In a series of personal photo albums with commentary, Grmek wrote: "Every living species has particular and typical means of survival. The elephant, its strength; the bird, its ability to fly; and the bacterium, its adaptability. Man has its intelligence. We understand things only to the extent that they can be useful to our survival and to exert domination over other beings and the natural world. Metaphysical questions go beyond our understanding because it was not constituted to face them." Cited in Lambrichs, "Un intellectuel européen engagé," 106 [my translation].

25. Ivica Vučak, "In Memoriam Mirko Dražen Grmek (1924–2000)," *Croatian Medical Journal* 41 (2000): 213.

26. His son's name was Smiljan Antun Grmek; he died in 1998.

27. On Grmek's medical internship, see Tatjana Buklijas, "Dalmacija u djelu Mirka Dražena Grmeka (1924–2000)," *Građa i prilozi za povijest Dalmacije* 16 (2000); Vučak, "In Memoriam."

28. Mirko D. Grmek, "Dernière Interview," in *La guerre comme maladie sociale*, 233 [my translation].

29. Grmek also often referred to Sigerist as one of his "masters." See, Mirko D. Grmek, "Review of H. E. Sigerist *Autobiographical Writings* and *A Bibliography of the Writings of Henry Sigerist*," *Isis* 59 (1968): 464.

30. In his 1971 CNRS report, Grmek explained that his "professional vocation was born out of the need to escape the servitude of specialization." See Mirko D. Grmek, "Titres et travaux" (unpublished manuscript, 1971), GRM/BE/1, p. 5, Grmek Papers, IMEC, Caen, France [my translation]. As to Sigerist, he said: "I think that what determined my career was that I steadily refused to specialize. All my teachers wanted to make me a specialist. . . . But my interests were very broad." Henry E. Sigerist, *Landmarks in the History of Hygiene* (London: Oxford University Press, 1956), vi. Grmek knew of Sigerist's opinion about specialization and even emphasizes it in his 1968 review. See Grmek, "Review of H. E. Sigerist," 463.

31. On this research published in English, see Mirko D. Grmek, "Ancient Slavic Medicine," *Journal of the History of Medicine and Allied Sciences* 14, no. 1 (1959).

32. For an analysis of Grmek's works on Dubrovnik scientists see, Tatjana Buklijas, "Dubrovnik in the Work of Mirko Dražen Grmek," *Dubrovkik Annals* 5 (2001).

33. "I was perfectly delighted to hear that your Academy has created a special Division for the History of Medicine. This will give Dr. Grmek an opportunity to develop his studies and I expect a great deal from him. His approach to the subject is similar to mine and I am sure that the archives of your country still have very important source materials that have not been investigated yet." Henry Sigerist to Andrija Štampar, June 4, 1953, GRM2.Da-10, Grmek Papers, IMEC, Caen, France.

34. Mirko D. Grmek, "The Study and Teaching of the History of Medicine in Yugoslavia," *Liječnički vjesnik* 84 (1962): 4.

35. On Štampar, see Patrick Zylberman, "Fewer Parallels than Antitheses: René Sand and Andrija Štampar on Social Medicine, 1919–1951," *Social History of Medicine* 17 (2004); Želijko Dugac, "New Public Health for a New State: Interwar Public Health in the Kingdom of Serbs, Croats, and Slovenes and the Rockefeller Foundation," in *Facing Illness in Troubled Times: Health in Europe in the Interwar Years, 1918–1939*, ed. Iris Borowy and Wolf D. Gruner (Frankfurt, Germany: Lang, 2005). Štampar's papers in public health can be found in *Serving the Cause of Public Health: Selected Papers from Andrija Štampar*, ed. Mirko D. Grmek (Zagreb: School of Public Health, 1966).

36. See Andrija Štampar, "Some Comments on the Law for the Protection of National Health," in *The History of East-Central European Eugenics, 1900–1945*, ed. Marius Turda (London: Bloomsbury, 2015). Štampar's views were echoed by Lujo Thaller, who, for instance, when discussing abortion in 1920, proposed to entrust physicians "with the authority to decide on biological matters related to the future of the population." Cited in Vjera Duić, "Overview," in *The History of East-Central European Eugenics, 1900–1945*, 424. For a recent analysis of Štampar's eugenicist views, see Martin Kuhar, "'From an Impure Source,

All Is Impure': The Rise and Fall of Andrija Štampar's Public Health Eugenics in Yugoslavia," *Social History of Medicine* 30 (2016), and on Štampar's organicist approach to society, see Stella Fatović-Ferenčić, "'Society as Organism': Metaphors as Departure Point of Andrija Štampar's Health Ideology," *Croatian Journal of Medical History* 49 (2008).

37. The WHO definition of *health* as "a state of complete physical, mental and social well-being and not merely the absence of disease or infirmity" is a shorter version of Štampar's proposal to define *health* "not only as the absence of infirmity and disease but also a state of physical and mental well-being and fitness resulting from positive factors, such as adequate feeding, housing and training." Cited in Kuhar, "'From an Impure Source,'" 92. Grmek also refuses to reduce the nature of health as "the calculation of one's disease and infirmities" and goes on to include in the definition of *health* the capacity for an individual to remain in a "state of balance with his surrounding world." Mirko D. Grmek, "La gérontologie d'hier à aujourd'hui," in *Deux cents ans de gérontologie* (Paris: La Société de gérontologie d'Ile de France, 1990), 10 [my translation].

38. Vučak, "In Memoriam," 214.

39. Mirko D. Grmek, "Medicinska škole u Dalmaciji u vrijeme francuske vladavine, 1806–1813" [Medical schools in Dalmatia under the French protectorate, 1806–1813] (PhD diss., University of Zagreb, 1958).

40. Grmek, "Dernier Interview," 234.

41. Henry Sigerist to Richard Shryock, October 8, 1952, in *Correspondence Henry E. Sigerist to Richard H. Shryock (1933–1956)*, ed. Marcel H. Bickel (Bern: self-pub., 2012), 133, http://www.img.unibe.ch/unibe/portal/fak_medizin/ ber_vkhum/inst_medhist/content/e40437/e40444/e153944/section154575/ files154581/CorrespondenceHenryE.Sigerist-RichardH.Shryock_ger.pdf. This comment from Sigerist must be qualified, as it seems to imply there was no history of medicine in former Yugoslavia prior to the creation of Grmek's institute. However, the history of medicine had been taught by Thaller (who had a chair) since 1927. In 1946, the history of medicine was mainly taught at the Zagreb Medical School by Lavoslav Glesigner. See Grmek, "Study and Teaching."

42. "I was interested in your suggestion about the application of Dr. Grmek of Zagreb for a possible fellowship here, and I also received a letter from Professor Štampar in the same connection. It would be interesting in many ways to have a Yugoslav scholar resident in the Institute. But next year we shall be able to offer only one, or at most, two fellowships; and the announcement of these, including the statement that they are available only to Americans and Canadians, has already been widely circulated in the American journals. Under these circumstances I have decided, with some regrets, that it would not be wise to change this rule in this special instance." —Richard Shryock to Henry Sigerist, October 24, 1952, in *Correspondence Sigerist to Shryock*, 134.

43. Grmek writes: "Early in 1954, I qualified as an external lecturer in the history of medicine with a thesis on the life and works on the medieval medical

writer Dominiko of Dubrovnik (1558–1613), and during the same year I was appointed lecturer on the subject 'An Introduction to Medicine.'" Grmek, "Study and Teaching," 7.

44. Commenting further on the particularities of his lectures, Grmek says he tried "to explain to the students the meaning of basic medical ideas and by the same method to point out the principles of medical deontology and the problems of the social application of medicine (the link between disease and poverty, the system for developing medical services, the development of social insurance, international health cooperation, etc.)." Grmek, "Study and Teaching," 7. Photographer Milan Pavić took a picture of Grmek's class in 1957. See GRM 123, album 7 (1957–1958), Grmek Papers, IMEC, Caen, France.

45. Grmek, "Titres et travaux," 6 [my translation].

46. Grmek, 18.

47. This monograph was followed by other contributions on the history of gerontology and the relation between ageing and death. See, in particular, Mirko D. Grmek, "Le vieillissement et la mort," in *Biologie*, ed. J. Rostand and A. Tétry (Paris: Encyclopédie de la Pléiade, 1965).

48. "Definitions," in Mirko D. Grmek, *On Ageing and Old Age: Basic Problems and Historic Aspects of Gerontology and Geriatrics* (The Hague: W. Junk, 1958), 4.

49. It was Jean Théodoridès who introduced Grmek to Danièle Guinot, whom he married in 1963. Although she stayed somewhat in the background of Grmek's fame, Danièle—a zoologist at the Museum d'histoire naturelle in Paris—deserves credit for proofreading his papers and books in French from the late 1950s until the late 1980s. They officially divorced in 1977.

50. On the historiography of the Annales School, see Toby Gelfand, "The Annales and Medical Historiography: *Bilan et perspectives*," in *Problems and Methods in the History of Medicine*, ed. Roy Porter and Andrew Wear (London: Croom-Helm, 1987).

51. Mirko D. Grmek, *Cataglogue des manuscrits de Claude Bernard. Avec la bibliographie de ses travaux imprimés et des études sur son œuvre*, with a foreword by M. Bataillon and E. Wolff and an introduction by L. Delhoume and P. Huard (Paris: Masson, 1967).

52. Grmek, "Titres et travaux," 6, [my translation]. The term *epistemology* should be understood in the context of French history and philosophy of science, where it refers to the process of generating scientific knowledge, not to theories of knowledge. On this distinction, see Hans-Jörg Rheinberger, *On Historicizing Epistemology*, trans. David Fernbach (Stanford, Calif.: Stanford University Press, 2010), 2–3.

53. Grmek, "Titres et travaux," 6 [my translation]. Grmek obtained French citizenship in 1967.

54. On Grmek and the tradition of "disease ecology," see Pierre-Olivier Méthot, "Le concept de pathocénose chez Mirko Grmek: une réflexion évolutionniste sur l'écologie des maladies?" in Coste, Fantini, and Lambrichs, *Le concept de pathocénose de Mirko Grmek*.

55. Grmek's concept of pathocenosis draws from many sources but its parasitological origins have often been overlooked. Although no correspondence between them was found, Grmek was acquainted with the work of Russian parasitologist Evgeny Pavlovsky (1884–1965) already in 1963. Building on the concept of biocenose, the glossary of Pavlovsky's book *Natural Nidality of Transmissible Diseases*, translated into English in 1966—the year Grmek introduced the notion of pathocenosis in England and in France—contains several technical terms such as *biogeocenose, biocenology,* and, especially, *biopathocenose,* which may have inspired Grmek's neologism. See Evgeny N. Pavlovsky, *Natural Nidality of Transmissible Diseases with Special Reference to the Landscape Epidemiology of Zooanthroponoses,* trans. Frederick K. Plous Jr. (Urbana: University of Illinois Press, 1966), 247. On the history of the concept of pathocenosis, see Pierre-Olivier Méthot, "De la pathocénose aux maladies émergentes: production, circulation et transformation conceptuelles," in *Médecine, science, histoire. Le legs de Mirko Grmek,* ed. Pierre-Olivier Méthot (Paris: Matériologiques, forthcoming).

56. Mirko D. Grmek, "Discussion on Medicine and Culture," in *Medicine and Culture,* ed. Frederick N. L. Poynter (London: Wellcome Institute, 1969), 49.

57. Jacques Félician, Pierre Ginésy, Geneviève Krick, Jean-Paul Liauzu, and Jacques Saliba, "Entretien avec Mirko Grmek," *Césure* (1995): 230 [my translation].

58. Chapter 1, this volume.

59. Chapter 4, this volume.

60. Grmek's first visit to the United States was in August–September 1962 on the occasion of the 10th International Congress of the History of Science at Cornell University, Ithaca. Between 1967 and 1971, he was visiting associate professor at the University of California–Berkeley (1967), visiting lecturer at the Harvard Medical School (1969), and visiting professor at the California State College in Los Angeles (1971).

61. Grmek, "Introductory Lecture," 7–8.

62. Roger Hahn, "Berkeley's History of Science Dinner Club: A Chronicle of Fifty Years of Activity," *Isis* 90 (1999): 189.

63. Grmek knew the Hungarian biochemist Szent-Györgyi during a workshop he organized for Claude Bernard's centenary earlier in June 1965 at the Singer-Polignac Foundation in France.

64. For a report on the workshop, see Mirko D. Grmek, "Aux États-Unis, un colloque international sur les rapports entre la Biologie, l'Histoire et la Philosophie, Denver, 27 novembre-2 décembre 1967," *Revue d'histoire des sciences* 21 (1968). A photo from a local newspaper showing Szent-Györgyi, Ayala, Taton, and Grmek is available in Grmek's papers, GRM 126, album 13 (1967–1968), p. 8, Grmek Papers, IMEC, Caen, France.

65. GRM 126, album 13 (1967–1968), p. 89, Grmek Papers, IMEC, Caen, France.

66. Charles D. O'Malley, ed., *The History of Medical Education: An International Symposium Held February 5–9, 1968* (Los Angeles: University of California Press, 1970).

67. "Dear Dr. Grmek, my colleague, Charles Gillispie, has told me of his conversation with you in New York and has suggested that you might be able to give a talk to our staff and graduate students when you return to the United States in February. I should be particularly glad if we were able to make the necessary arrangements, for I have been hearing about you, both from Charles [Gillispie] and from Roger Hahn, for a very long time and should be delighted for an opportunity to make your acquaintance." Thomas Kuhn to Mirko D. Grmek, December 13, 1967, GRM2.Da-06, Grmek Papers, IMEC, Caen, France.

68. The manuscript of Grmek's "The Search for a General Theory of Ageing" is available at GRM 40, Grmek Papers, IMEC, Caen, France.

69. Chapter 5, this volume.

70. Grmek, "Titres et travaux," 19 [my translation].

71. The other lecturers were I. Bernard Cohen (1914–2003), James A. Shannon (1904–1994), Allen G. Debus (1926–2009), and Charles O'Malley. Grmek's talk was titled "Hippocrates and the Enigma of Life." This conference remained unpublished and was found in neither Grmek's papers in Caen nor Debus's, who chaired the organization of the lectures in Chicago.

72. GRM 127, Album 15 (1970–1973), p. 5, Grmek Papers, IMEC, Caen, France.

73. "Dear Dr. Grmek, I was very happy to hear through Dr. Chapin that you will be able to accept our invitation to give a Biophysics Seminar here on Monday, May 17th at 4 pm, with the title 'Descartes as a Biophysicist.'" Max Delbrück to Mirko D. Grmek, May 7, 1971, GRM2.Da-03, Grmek Papers, IMEC, Caen, France.

74. Mirko D. Grmek, GRM 127, Album 15 (1970–1973), p. 5, Grmek Papers, IMEC, Caen, France.

75. Mirko D. Grmek, Robert S. Cohen, and Guido Cimino, eds., *On Scientific Discovery: The Erice Lectures 1977* (Boston: Reidel, 1981).

76. Here I depart from Duffin's perspective when she writes that "until the 1980s, [Grmek's] reputation in North America was modest and centered on philological contributions." Jacalyn Duffin, "In Memoriam: Mirko Dražen Grmek," *Bulletin of the History of Medicine* 74 (2000): 563.

77. For a recent essay on the relationship between the history of science and the history of medicine, see Staffan Müller-Wille, "History of Science and Medicine," in *The Oxford Handbook of the History of Medicine*, ed. Mark Jackson (Oxford: Oxford University Press, 2011).

78. Later renamed the Centre Alexandre Koyré.

79. On the history of these two institutes and their specific intellectual orientations, see Claire Salomon Bayet, "L'histoire des sciences et des techniques," in *L'histoire et le métier d'historien en France 1945–1995*, ed. François Bédarida

(Paris: Maison des sciences de l'homme, 1995); Pietro Redondi, "Ernest Coumet et l'histoire de l'histoire des sciences," *Revue de synthèse* 4 (2001); and Jean Gayon, "L'Institut d'histoire des sciences," *Cahiers Gaston Bachelard* 14 (2016).

80. In addition to giving public lectures at the Palais de la découverte and teaching on the history of biological thought at Canguilhem's Institute in 1963–1964, Grmek contributed one article in *Mélanges Alexandre Koyré*, a collective volume, and three articles in Taton's encyclopedia of history of science. See Mirko D. Grmek, "La conception de la maladie et de la santé chez Claude Bernard," in *Mélanges Alexandre Koyré*, vol. 2, *L'Aventure de la science* (Paris: Hermann, 1964); Mirko D. Grmek, "La science chez les Slaves du Moyen-Age," in *La science antique et médiévale. Des origines à 1450*, ed. René Taton (Paris: Presses universitaires de France, 1966); and "L'étude du corps humain" and "L'art de guérir" (with Paul Delaunay), in *La science moderne. De 1450 à 1800*, ed. René Taton (Paris: Presses universitaires de France, 1969).

81. "This book would have never seen the light of day without the initiative of Fernand Braudel at a time now distant, or without the encouragements which that famous historian of Mediterranean civilization was willing to lavish upon me." Grmek, *Diseases in the Ancient Greek World*, xii. On Grmek and Braudel, see Jon Arrizabalaga, "History of Disease and the Longue Durée," *History and Philosophy of the Life Sciences* 27 (2005).

82. See Mirko D. Grmek, "L'émergence de la médecine scientifique en France sous le règne de Louis XIV," *Medizinhistorishes Journal* 11 (1976): 274; Mirko D. Grmek, *La première révolution biologique. Réflexions sur la physiologie et la médecine au XVIIᵉ siècle* (Paris: Payot, 1990). Grmek has long suggested that the "fundamental characteristic of sciences" is that "their progress is cumulative," although "this accumulation is not simply a matter of addition." Grmek, "Prolégomènes," 145 [my translation].

83. Mirko D. Grmek, introduction to *Western Medical Thought from Antiquity to the Middle Ages*, ed. Mirko D. Grmek, coord. Bernardino Fantini, trans. Antony Shugaar (Cambridge, Mass.: Harvard University Press, 1998), 20.

84. According to Grmek, "continuity and rupture coexist and any historical venture based on an exclusive mythicizing of one of these two complementary aspects would seem doomed in advance." Chapter 7, this volume.

85. Cited in Grmek, "Médecine et épistémologie," 3 [my translation]. "I would not have considered defending a thesis if Professor Georges Canguilhem had not encouraged me to do so. I admire the work of the director of the Institut d'Histoire des Sciences at the Sorbonne, and this is why I am even more honored of the support he always gave me." Mirko D. Grmek, *Raisonnement expérimental et recherches toxicologiques chez Claude Bernard* (Geneva: Droz, 1973), x [my translation].

86. See Alexandre Klein, "Quelle place pour Mirko Grmek, élève de Georges Canguilhem, dans l'historiographie médicale française?," in *Médecine, science, histoire. Le legs de Mirko Grmek*, ed. Pierre-Olivier Méthot (Paris: Matériologiques, forthcoming).

87. See Chapter 7, this volume. Mirko D. Grmek, "Définition du domaine propre de l'histoire des sciences et considérations sur ses rapports avec la philosophie des sciences," *History and Philosophy of the Life Sciences* 1 (1979): 5.

88. Grmek criticized philosophers of science, among whom, he said, one finds some "brilliant thinkers," "talented reasoners," and "great masters of discourses," for being "invasive" when they assign themselves the task of "dictating laws to the practice of science and a *priori* results to historical enquiries." Mirko D. Grmek, "Définition du domaine propre," 7 [my translation].

89. Grmek, "Introductory Lecture," 8.

90. Grmek, introduction, 17–18. Against "philosophical history" of science, by the end of his career Roger defended the value of a "historian history" of science (*histoire historienne*) and, like Koyré, the need to "understand the past in its own terms." See Jacques Roger, "Pour une histoire historienne des sciences," in *Pour une histoire des sciences à part entière*, ed. Claude Blanckaert (Paris: Albin Michel, 1995), 50–54, 66 [my translation].

91. Mirko D. Grmek, "Épidémiologie de la peste et histoire démographique et sociale," in *L'histoire des sciences et des techniques doit-elle intéresser les historiens? Colloque organisé par la Société française d'histoire des sciences et des techniques, 8 et 9 mai 1981* (Paris: Société française d'histoire des sciences et des techniques, 1982), 177 [my translation].

92. Georges Canguilhem (president), François Dagognet, Pierre Huard, and Charles Kayser (1899–1981) were on the thesis committee.

93. Chapter 7, this volume.

94. Chapter 7, this volume.

95. Gerald L. Geison, *The Private Science of Louis Pasteur* (Princeton, N.J.: Princeton University Press, 1995); Frederic L. Holmes, Jürgen Renn, and Hans-Jörg Rheinberger, eds., *Reworking the Bench: Research Notebooks in the History of Science* (Dordrecht, Netherlands: Kluver, 2003); Hans-Jörg Rheinberger, "History of Science and the Practice of Experiment," *History and Philosophy of the Life Sciences* 23 (2001). See also the Foreword to this volume.

96. See Duffin, "In Memoriam."

97. First secretary and later president of the Académie internationale d'histoire des sciences (1981–1985), Grmek was also editor-in-chief of the *Archives Internationales d'histoire des sciences*, the official journal of the academy (1965–1970).

98. The journal is now published by Springer and is currently being edited by Staffan Müller-Wille from the University of Exeter, UK.

99. The school continues to take place on a biannual basis. http://www .ischiasummerschool.org. On the history of the Stazione Zoologica, see Christiane Groeben, "Stazione Zoologica Anton Dohrn," *eLS* (2013), https://doi .org/10.1002/9780470015902.a0024932.

100. Inspired by the success of the Ischia School, American historian of biology Garland E. Allen (b. 1936) invited Grmek to guest lecture at the inaugural edition of the History of Biology Seminar at the Marine Biological Laboratory

in Woods Hole during the summer of 1987, insisting that Grmek's participation would provide "an element of continuity between Ischia and Woods Hole," which was seen as a "sister course." Garland E. Allen to Mirko D. Grmek, October 31, 1986, GRM2.Da-01, Grmek Papers, IMEC, Caen, France. The list of seminars that ran from 1987 until 2017 can be found at the History of Biology Seminars at the MBL page at Arizona State University's Center for Biology and Society, https://cbs.asu.edu/mbl.

101. Grmek, "Prolégomènes," 144.

102. Mirko D. Grmek, "Histoire des sciences et psychogenèse," *Cahiers de la Fondation Archives Jean Piaget* 4 (1983): 29 [my translation].

103. See Chapter 7, this volume.

104. Grmek, "Définition du domaine propre," 11.

105. Grmek, "Discours de Rome," 2 [my translation].

106. This talk was given during the conference held in Paris in 1969 and organized by Canguilhem, on the occasion of the bicentennial anniversary of French paleontologist and comparative anatomist Georges Cuvier (1769–1832). Foucault's opening address was followed by commentaries from François Dagognet, Camille Limoges, François Courtès, and Yvette Conry. Grmek, who chaired the discussion, raised several questions about Foucault's argument. This paper and the commentaries were recently translated into English by Lynne Huffer. See Michel Foucault, "Cuvier's Situation in the History of Biology," *Foucault Studies* 22 (2017). Grmek's cited comments are on page 230.

107. O. Ouvry, D. Marcelli, and Y. Edel, "Le sida remis dans son histoire. Entretien avec le Pr Grmek," *Nervure* 3 (1990): 65.

108. This connection between Grmek and Daremberg was pointed out by Jean-François Braunstein in a recent interview. See Jean-François Braunstein, "Entretien avec Jean-François Braunstein," *Intelligere, Revista de História Intelectual* 2 (2016): 182. For an analysis of this connection, see especially Klein, "Quelle place pour Mirko D. Grmek."

109. See Georges Canguilhem, "The Object of the History of Sciences," in *Continental Philosophy of Science*, ed. Garry Gutting (Blackwell: Malden, 2005), 203. Canguilhem criticizes internalist historians who, when treating historical facts like scientific facts, that is, "facts of which one cannot write the history without a theory," are at once "aligning the history of science with science when it is a matter of the relation of knowledge to its object." Canguilhem, 202. The position described by Canguilhem seems to be Grmek's when he claims that in order to interpret a historical fact, "the historian relies on a 'historical theory,' just as a physicist or a biologist makes use of a theory to interpret an 'experimental fact.'" Grmek, "Introduction," 19.

110. Grmek, "Prolégomènes," 144.

111. Roy Porter, "The Patient's View: Doing History from Below," *Theory and Society* 14, no. 2 (1985). On Porter, see the recent essays collected in Alexandra Bacopoulos-Viau and Aude Fauvel, eds., "Tales from the Asylum: Patient Narratives and the (De)Construction of Psychiatry," special issue, *Medical History* 60, no. 1 (2016).

112. Having an MD to practice medical history was a requirement put forward by Sigerist, for instance. Citing his famous predecessors who were practicing medical doctors, he goes on to define the historian of medicine as "a physician, trained in the research methods of history, who takes an active part in the life of his time and is in close touch with the medical problems of his time." Henry E. Sigerist, *A History of Medicine*, vol. 1, *Primitive and Archaic Medicine* (New York: Oxford University Press, 1951), 31.

113. Charles Rosenberg, "Disease in History: Frames and Framers," *The Milbank Quarterly* 67, no. S1 (1989). See the comments on Rosenberg by Wilson, "On the History of Disease-Concepts." Mirko D. Grmek, "La révolution biomédicale du XXᵉ siècle," in *Histoire de la pensée médicale en Occident*, vol. III, ed. Mirko D. Grmek (Paris: Seuil, 1999), 334 [my translation].

114. Grmek, "Introduction," 18.

115. Jacalyn Duffin was the Hannah Chair in the History of Medicine at York University, Canada, from 1987 to 2016, where she is now an emeritus professor.

116. Mirko D. Grmek to Jacalyn Duffin, November 22, 1986, GRM2.Da-03, Grmek Papers, IMEC, Caen, France [my translation]. Grmek's response was prompted by an earlier letter from Duffin in which she describes her experience during the congress of the American Association for the History of Medicine at Rochester in 1986. Duffin mentions she was labeled as an "internalist" by some of Charles Rosenberg's graduate students. Not "swayed by the exaggerated social trends," she enjoyed recounting she was Grmek's student: "I found it quite amusing and conversation provoking," Duffin writes, "to announce that I was *your* student. Reactions from disbelief to uncomprehending blank stares." Jacalyn Duffin to Mirko D. Grmek, May 5, 1986, GRM2.Da-03, Grmek Papers, IMEC, Caen, France. In a 1989 letter to her former thesis advisor, Ana Cecilia Rodríguez de Romo (b. 1954), who was then on a scholarship at the Institute for the History of Medicine at Johns Hopkins, comments on the status of medical history, echoing Grmek's teaching: "It seems that what dominates here in the United States is what they called 'Social History.' . . . One also has to be a 'specialist.' It is not possible to touch on every subject." Ana Cecilia Rodríguez de Romo to Mirko D. Grmek, Baltimore, Md., January 24, 1989, GRM2.Da-09, Grmek Papers, IMEC, Caen, France [my translation].

117. Mirko D. Grmek, introduction to *Histoire de la pensée médicale en Occident*, vol.1, *Antiquité et Moyen Age*, ed. Mirko D. Grmek, in collaboration with Bernardino Fantini (Paris: Seuil, 1995), 23.

118. Throughout his career, Grmek contributed to the rise of paleopathology in France by his teaching and by introducing a new scientific classification of bone tumors. See, Mirko D. Grmek, "La paléopathologie des tumeurs osseuses malignes. Proposition d'une classification à l'usage de l'ostéo-archéologie, revue des exemples publiés et présentations de deux cas inédits," *Histoire des sciences médicales* 9 (1975). As paleopathology is fast developing, thanks to next generation sequencing apparatuses, historians claim that some of the problems of retrospective diagnosis could be overcome and, in the process, have rediscovered Grmek's work. See Monica H. Green, "The Value of Historical Perspective,"

in *Research Companion to the Globalization of Health*, ed. Ted Schrecker (Aldershot, England: Ashgate, 2012), 35. For critiques of retrospective diagnosis, see Jon Arrizabalaga, "Problematizing Retrospective Diagnosis in the History of Disease," *Asclepio* 54 (2002).

119. Mirko D. Grmek, "La réalité nosologique au temps d'Hippocrate," in *La collection hippocratique et son rôle dans l'histoire de la médecine (Strasbourg, 1972)*, ed. Mirko D. Grmek (Leiden, Netherlands: Brill, 1975), 239 [my translation].

120. Grmek was particularly impressed by the work of the British American anthropologist John L. Angel (1915–1986), who conducted fieldworks in Greece before World War I, and would later describe him as the "personification" of paleopathology. Grmek, *Diseases in the Ancient Greek World*, 55.

121. "To us it seems entirely reasonable to assert the immutability of natural laws as an epistemological postulate, at least on the scale of human time and human space. As for the biological evolution of man, the scientific evidence currently available points strongly to its slowness. Accordingly, we infer that the regulatory mechanisms and pathological reactions of the human body on all levels of its organization are the same now as in the past." Grmek, *Diseases in the Ancient Greek World*, 8. Grmek first expressed this view in "La réalité nosologique," 248–49.

122. Gilberto Corbellini and Chiara Preti, "Towards an Evolutionary Historiography and Epistemology of Medicine: The Legacy of Mirko Grmek," *Medicina Nei Secoli* 20 (2008).

123. Frank Huisman, "The Dialectics of Understanding: On Genres and the Use of Debate in Medical History," *History and Philosophy of the Life Sciences* 27 (2005): 25.

124. Fernand Robert, "Un événement dans les études grecques: Le nouveau livre de Mirko D. Grmek," *Bulletin de l'Association Guillaume Budé* 2 (1984) [my translation]; Georges Canguilhem to Mirko D. Grmek, January 21, 1984, GRM2.Da-02, Grmek Papers, IMEC, Caen, France [my translation].

125. Shaw, "Grmek's Pathological Vision," 331.

126. Jones, "The Pathocenosis Moment," 7.

127. See Robert Sallares, *The Ecology of the Ancient Greek World* (Ithaca, N.Y.: Cornell University Press, 1991). See also Robert Sallares, "Pathocenosis: Ancient and Modern," *History and Philosophy of the Life Sciences* 27 (2005).

128. Roy Porter, "Review of *Histoire du sida: Début et origine d'une pandémie actuelle,*" *Medical History* 34 (1990), 458. In spite of divergent methodological approaches, Porter praises Grmek's book on AIDS and contributed a chapter ("Les stratégies thérapeutiques") in Grmek's second edited volume of the *Histoire de la pensée médicale*, published in 1997.

129. Mirko D. Grmek, "Sida et relations internationales. Entretien avec Mirko D. Grmek," *Politique internationale* 50 (1991): 362 [my translation].

130. "Entretien avec Mirko Grmek. Sida: Histoire d'une épidémie," *L'Histoire* 150 (1991).

131. Gerald M. Oppenheimer, "Review of Mirko D. Grmek, *History of AIDS: Emergence and Origin of a Modern Pandemic,*" *Isis* 83 (1992): 694.

132. Michaël Pollack, "Comptes rendus: *Histoire du sida. Début et origine d'une pandémie actuelle*," *Annales E.S.C.* 44 (1989).

133. Thomas W. Laqueur, "Viral Cultures," *The New Republic* (1991): 38.

134. Laqueur, 39.

135. Gourevitch, *Maladie et maladies*.

136. Danielle Gourevitch and Mirko D. Grmek, *Les maladies dans l'art antique* (Paris: Fayard, 1998).

137. Roger Hahn, "Sarton Medal Citation," *Isis* 83 (1992): 282. Grmek was also awarded two honorary doctorates, one from the University of Lausanne (philology) in 1987 and another from the University of Bologna (history and philosophy of science) in 1998.

138. Stephen Morse, email message to Pierre-Olivier Méthot, May 25, 2014.

139. See Patrick Zylberman, *Tempêtes microbiennes. Essai sur la politique de sécurité sanitaire dans le monde transatlantique* (Paris: Gallimard, 2013).

140. Speaking of AIDS, Grmek claims that "the current pandemic occurred because radical changes in human behavior interrupted the long-standing equilibrium between host and parasite." Mirko D. Grmek, *History of AIDS: Emergence and Origin of a Modern Pandemic*, trans. Russell C. Maulitz and Jacalyn Duffin (Princeton, N.J.: Princeton University Press, 1990), 196. See Mirko D. Grmek, "The Concept of Emerging Disease," translated and reprinted in this volume. For a discussion of Grmek's view on emerging infections, see Pierre-Olivier Méthot and Bernardino Fantini, "Medicine and Ecology: Historical and Critical Perspectives on the Concept of 'Emerging Disease,'" *Archives Internationales d'histoire des sciences* 64 (2014). For recent work on emerging infections, see Simon Carter, John L. Allen, Nick Bingham, and Steve Hinchliffe, *Pathological Lives: Disease, Space, and Biopolitics* (Chichester, England: Wiley, 2016).

141. Mirko D. Grmek, "Regard d'un historien sur les maladies émergentes," in *Proceedings of the XXth International Congress of History of Science (Liège, 20–26 July 1997)*, ed. Denis Buican and Denis Thieffry (Turnhout, Belgium: Brepols, 2002), 16 [my translation].

142. See Grmek, "Dernier Interview," 240–41.

143. The International Croatian Initiative was active between 1992 and 1998 and its members were from many Western countries, including the Netherlands, Germany, the United States, France, Canada, Italy, and Switzerland. For a list of the founding members of the association, see Lambrichs, "Un intellectuel européen engagé," 168.

144. Mirko D. Grmek, "L'ex-Yougoslavie: la guerre comme maladie sociale," in *La guerre comme maladie sociale*.

145. Mirko D. Grmek, Marc Gjidara, and Neven Simac, *Le nettoyage ethnique. Documents historiques sur une idéologie serbe* (Paris: Le Seuil, 1993). This book sparked a controversy upon its publication, although it was generally well reviewed in the French press. Some scholars accused the three authors (all of whom have Croatian origins) of misrepresenting the Serbian culture in attributing the practice of "ethnic cleansing" to Serbians and in grounding it in the history and the biology of the Serbian peoples more generally. Grmek, Gjidara,

and Simac filed a lawsuit for libel against French sociologist and historian Catherine Lutard in September 1993. Louise Lambrichs provides an account of the two trials in "Nettoyage ethnique: le procès," in Grmek, *La guerre comme maladie sociale*. For a description of these events from Lutard's point of view, see Catherine Lutard-Tavard, "Être à la barre, être accusé(e)," *Socio* 3 (2014). And for an analysis of the book and the history of "ethnic cleansing" more generally see Alice Krieg-Planque, *"Purification ethnique." Une formule et son histoire* (Paris: CNRS Éditions, 2003).

146. Mirko D. Grmek, "Un mémoricide," in *La guerre comme maladie sociale*.

147. Grmek, "L'ex-Yougoslavie," 101 [my translation].

148. Mirko D. Grmek, "Le concept de maladie," in Grmek, *La vie, les maladies et l'histoire*, 28 [my translation].

149. Grmek, "Le concept de maladie," 28 [my translation].

150. Comparing Grmek's concept of "social disease" with Štampar's organicist view is useful. Štampar once remarked that: "As an individual organism consists of cells, so does the society consists of individuals representing cells; as an individual organism can become ill, so can the society become ill." This passage is translated in Fatović-Ferenčić, "'Society as Organism,'" 711.

151. Holmes wrote the following letter two weeks before Grmek's death:

Your recollection of happy recent encounters with one another brought to my mind memories of earlier ones that also still mean much to me. I remember vividly the first time, in Banyuls, when you encouraged me about the paper I had read, and explained to me why I must see the Claude Bernard archives in Paris. Up until then I had worked only with published materials, and knew nothing about the importance of unpublished documents. I remember coming to your apartment in Paris, and how helpful you were introducing me to the documents. I also remember several very pleasant conversations over coffees on the sidewalk of Paris. I was, of course, a little bit intimidated, because you knew so much about so many things, and I felt I knew very little . . . I am particularly sad right now that there is no prospect that I can come to Paris for one last happy encounter.

Larry Holmes to Mirko D. Grmek, January 24, 2000, box 9, F. L. Holmes Papers, Yale University Library, New Haven, Conn.

152. Bernardino Fantini, "Obituary: Mirko Dražen Grmek (1924–2000)," *Medical History* 45 (2001): 276.

153. On Grmek's use of barbiturate drugs, see Louise L. Lambrichs, "Note sur la préhistoire de ce volume," in Coste, Fantini, and Lambrichs, *Le concept de pathocénose de Mirko Grmek*, 9–10.

154. Mirko D. Grmek, "Préliminaires d'une étude historique des maladies," *Annales: Économie, Sociétés, Civilisations* (1969). The first occurrence of the term *pathocenosis* I traced in Grmek's correspondence is a letter from him to Emmanuel Le Roy Ladurie (b. 1929) and Jean-Pierre Peter (b. 1933) in 1966, four months before he presented the concept publicly at the Symposium on Medicine and Culture in London. These scholars, who worked with Braudel

(Peter also worked with Foucault), were seeking Grmek's help in translating eighteenth-century medical terms into modern terminology. In his response, Grmek explained the problems with such a project and mentioned he was working on a book about the "historical changes in pathocenosis," a term he then defined as the "ensemble of diseases of a human group that is geographically limited." Mirko D. Grmek to Emmanuel Le Roy Ladurie, May 12, 1966, GRM2.Da-06, GRM 73, Grmek Papers, IMEC, Caen, France [my translation].

155. See Stephen S. Morse, ed., *Emerging Viruses* (Oxford: Oxford University Press, 1993).

156. Here, Grmek particularly considers the "trade-off" model between virulence and transmissibility in host-parasite interactions. On the history and epistemology of this model, see Samuel Alizon, A. Hurford, N. Mideo, and M. Van Baalen, "Virulence Evolution and the Trade-Off Hypothesis: History, Current State of Affairs, and the Future," *Journal of Evolutionary Biology* 22 (2009).

157. Grmek and Holmes have debated the significance of Bernard's astonishment in finding sugar in blood independent of alimentation over nearly forty years, both publicly and privately. See Frederic L. Holmes, *Claude Bernard and Animal Chemistry: The Emergence of a Scientist* (Cambridge, Mass.: Harvard University Press, 1974), 423–25. See also the correspondence between Grmek and Holmes kept at IMEC and Yale University's Frederic L. Holmes Papers.

158. For a comprehensive analysis of Grmek's studies of ageing, see Jérôme Brousseau, "La sénescence et la mort des individus sont-elles utiles à l'espèce? Regards sur la théorie évolutionniste du vieillissement de Mirko D. Grmek," in *Médecine, science, histoire. Le legs de Mirko Grmek* (Paris: Matériologiques, forthcoming).

159. Grmek, Cohen, and Cimino, introduction to *On Scientific Discovery*, 1.

160. Grmek, Cohen, and Cimino, 5.

1. PRELIMINARIES FOR A HISTORICAL STUDY OF DISEASES

1. Cf. Mirko D. Grmek, "Histoire des recherches sur les relations entre le génie et la maladie," *Revue d'Histoire des Sciences* 15 (1962).

2. See the instructive study by Albert Esser, *Cäsar und die julisch-claudischen Kaiser in biologisch-ärztlichen Blickfeld* (Leiden, Netherlands: Brill, 1958).

3. [In his *Pensées* (no. 162), French philosopher Blaise Pascal wrote: "Cleopatra's nose: had it been shorter, the whole aspect of the world would have been altered." See Blaise Pascal, *Pascal's Pensées*, trans. W. F. Trotter, introduction by T. S. Eliot (New York: Dutton, 1958), 48.]

4. Without a doubt, there exist certain critical situations where the health of a military or political leader can have important historical consequences. Concerning contemporary military leaders, see R. Hargreaves, "The Saving Quality," *Practitioner* 190 (1963). Concerning kings and important political personalities of the past, we cannot be too careful in passing judgment on the historical

importance of their illnesses. Unfortunately, at the end of the last century and during the present century, readers were drowned in anecdotal literature on such subjects, as by a real tidal wave. In France, a prolix coryphaeus was found in the person of Doctor Augustin Cabanès. He tried to justify his method (*Cf.* Augustin Cabanès, *L'histoire éclairée par la clinique* [Paris: Albin Michel, 1921]), but it was impossible for him to go beyond the research stages of "curiosities" and "historical medical indiscretions." And yet medicine and history can mutually and significantly enrich each other as so well shown by Pierre Huard, "La médecine et l'histoire," *Revue de synthèse* 37–39 (1965).

5. For bibliographic indications, see the fundamental monograph by Wilhelm Lange-Eichbaum, *Genie, Irrsinn und Ruhm*, 5th ed. (Munich: Reinhardt, 1961), the works of Ralph H. Major, *Disease and Destiny* (New York: Appleton, 1936), and by Judson B. Gilbert and Gordon E. Mestler, *Disease and Destiny: A Bibliography of Medical References to the Famous* (London: Dawsons, 1962).

6. Littré defines *pathological history* as a "science still in an embryonic state and whose development will be one of the intellectual developments of the century." [Here, Grmek cites an aphorism often credited to Littré but rarely referenced. It was cited, for example, by Auguste Brachet, a student of Littré, in his *Pathologie mentale des rois de France. Louis XI et ses ascendants* (Paris: Hachette, 1903), iii.]

7. A few interesting ideas on the historical role of epidemics can be found in J. P. Granger's article "The Plague as a Factor in History," *Glasgow Medical Journal* 77 (1912). For the historical incidence of the "great" epidemics, see, for example, Heinrich Haeser, *Lehrbuch der Geschichte der Medicin und der epidemischen Krankheiten* (Jena, Germany: Mauke, 1882); Albert Colnat, *Les épidémies et l'histoire* (Paris: Le François, 1937); Erich Martini, *Die Wege der Seuchen* (Stuttgart, Germany: Enke, 1954); Ronald Hare, *Pomp and Pestilence* (London: Gollancz, 1954); Ralph H. Major offers a summery on the relations between the wars and epidemics ("War and Disease," *Journal of Laboratory and Clinical Medicine* 28 [1943]), whereas the military importance of syphilis is exposed with much verve in Hans Zinsser's book, *Rats, Lice, and History* (Boston: Little Brown, 1935). Concerning yellow fever, see Henry R. Carter, *Yellow Fever: An Epidemiological and Historical Study of Its Place and Origin* (Baltimore, Md.: Williams and Wilkins, 1931). Concerning the French colonial experience, see N. P. Gilbert, *Histoire médicale de l'armée française à Saint-Domingue en 1802* (Paris: Gabon et Cie, 1803); Paul Brau, *Trois siècles de médecine coloniale française* (Paris: Vigot, 1931); and Francisco Guerra, "The Influence of Disease on Race, Logistics and Colonization in the Antilles," *Journal of Tropical Medicine and Hygiene* 69 (1966).

8. Charles G. Gruner, *Morborum antiquitates* (Vratislavia, Bohemia: Korn, 1774).

9. August Hirsch, *Handbuch der historisch-geographischen Pathologie*, 3 vols. (Stuttgart, Germany: Enke, 1881–1886)

10. Charles Creighton, *A History of Epidemics in Britain* (London: Cambridge University Press, 1891). A very interesting review was published at the time of

the new edition in 1965 by R. S. Robert, "Epidemics and Social History: An Essay Review," *Medical History* 12 (1968).

11. Alfonso Corradi, *Annali delle epidemie occorse in Italia delle prime memorie fino al 1850*, 4 vols. (Bologna: Gamberini and Parmeggiani, 1865–1880).

12. Georg Sticker, *Abhandlungen aus der Seuchengeschichte und Seuchenlehre*, 3 vols. (Giessen, Germany: Töpelmann, 1908–1912).

13. John D. Rolleston, *The History of the Acute Exanthemata* (London: Heinemann, 1937).

14. Maxime Laignel-Lavastine, *Histoire générale de la médecine, de la pharmacie, de l'art dentaire et de l'art vétérinaire*, 3 vols. (Paris: Albin Michel, 1934–1949). This work contains several chapters dedicated to the history of different diseases; the value of these chapters is rather uneven.

15. Adalberto Pazzini and Aroldo Baffoni, *Storia delle malattie* (Rome: Clinica nuova, 1950).

16. Walter R. Bett, *The History and Conquest of Common Diseases* (Norman: University of Oklahoma Press, 1954).

17. Erwin H. Ackerknecht, *Geschichte und Geographie der wichtigsten Krankheiten* (Stuttgart, Germany: Inke, 1963).

18. Henschen Folke, *The History and Geography of Diseases* (New York: Delacorte, 1966).

19. Leonhard L. Finke, *Versuch einer allgemeinen medicinisch-praktischen Geographie*, 3 vol. (Leipzig, Germany: Weidmann, 1792–1795).

20. Cf. Mirko D. Grmek, "Géographie médicale et histoire des civilisations," *Annales: Économies, Sociétés, Civilisations* 28 (1963).

21. Jacques M. May, *Ecology of Human Disease* (New York: M. D. Publications, 1959); *Studies in Disease Ecology* (New York: Hafner, 1961).

22. [Wagner-Jauregg (1857–1940) thought that infectious diseases could be used to cure patients who were mentally ill, a technique known as fever therapy.]

23. Albert B. Sabin, "Nature of Inherited Resistance to Viruses," *Proceedings of the National Academy of Science* 38 (1952).

24. [On Th. Smith and the host-parasite hypothesis, see Pierre-Olivier Méthot, "Why Do Parasites Harm Their Host? On the Origin and Legacy of Theobald Smith's 'Law of Declining Virulence'—1900–1980," *History and Philosophy of the Life Sciences* 34 (2012).]

25. [*Inapparent infection* is a term coined by French microbiologist Charles Nicolle (1866–1936) to designate a silent form of infectious process that goes unnoticed by clinicians.]

26. René J. Dubos, *The White Plague: Tuberculosis, Man and Society* (Boston: Little, 1952).

27. René J. Dubos and James G. Hirsch, *Bacterial and Mycotic Infections of Man*, 4th ed. (Philadelphia: Lippincott, 1965).

28. John Lowe and F. McNulty, "Tuberculosis and Leprosy: Immunological Studies in Healthy Persons," *British Medical Journal* 2 (1953).

29. H. Floch, "La réaction de Mitsuda rendue positive par une primo-

infection tuberculeuse est-elle accompagnée d'une immunité relative anti-lépreuse?" *Bulletin de la Société de Pathologie Exotique* 47 (1954).

30. Roland Chaussinand, "Tuberculose et lèpre, maladies antagoniques. Éviction de la lèpre par la tuberculose," *International Journal of Leprosy* 16 (1948).

31. In particular, see the Japanese experiences: Ken Yanagisawa, "The Effect of BCG Vaccination upon Occurrence of Leprosy in Nursery Children," *International Journal of Leprosy* 26 (1958).

32. This hypothesis was first expressed in the article Henry E. Sigerist, "Der Aussatz auf den Hawaiischen Inseln," *Verhandlungen der Schweizer. Naturforschenden Gesellschaft* (1932).

33. The bibliography on the "Black Death" is huge. Classical works include: Justus C. F. Hecker, *Der schwarze Tod im vierzehnten Jahrhundert* (Berlin: Verlag, 1832); Cardinal F. A. Gasquet, *The Black Death of 1348 and 1349* (London: Bell, 1893); A. Phillippe, *Histoire de la peste noire (1348–1350) d'après des documents inédits* (Paris, 1853); J. Nohl, *Der Schwarze Tod. Ein Chronik der Pest 1348 bis 1720* (Potsdam, Germany: Kiepenheuer, 1924). For sources see A. Coville, "Écrits contemporains sur la peste de 1348 à 1350," *Histoire littéraire de la France* 37 (1928). For the social and economic impacts, I draw your attention to the following in particular: H. B. Allyn, "The Black Death: Its Social and Economic Results," *Annals of Medical History* 7 (1925); Y. Renouard, "Conséquences et intérêt démographique de la peste noire de 1348," *Population* 3 (1948); Elisabeth Carpentier, "Autour de la peste noire: famines et épidémies dans l'histoire du XIVᵉ siècle," *Annales: Économies, Sociétés, Civilisations* 17 (1962); Elisabeth Carpentier, *Une ville devant la peste, Orvieto et la peste noire de 1348* (Paris: S.E.V.P.E.N., 1963); William M. Bowsky, "The Impact of the Black Death upon Sienese Government and Society," *Speculum* 39 (1964); Josiah C. Russell, "Effects of Pestilence and Plague, 1315–1385," *Comparative Studies in Society and History* 8 (1966); Sylvia L. Thrupp, "Plague Effects in Medieval Europe," *Comparative Studies in Society and History* 8 (1966). In spite of the great number of studies of this type, their results have yet to be included in general history manuals. As per William L. Langer ("The Next Assignment," *The American Historical Review* 63 [1958]), these avenues of research should be among the most urgent tasks of present-day historiography.

34. B. E. Holsendorf, "The Rat and Ratproof Construction of Buildings," suppl. no. 131 (Washington, D.C.: U.S. Public Health Service, 1937).

35. For an update, see Leonard Fabian Hirst, *The Conquest of Plague: A Study of the Evolution of Epidemiology* (Oxford: Clarendon Press, 1953), and Georges Blanc, "La disparition de la peste et ses causes épidémiologiques," *Semaine des hôpitaux* 37 (1961).

36. Professor Henry H. Mollaret at the Pasteur Institute in Paris is presently conducting studies in this sense.

37. [In place of "chromosomal" Grmek should have written "genetic" abnormalities. I thank Jackie Duffin for pointing this out.]

38. Anthony C. Allison, "Protection Afforded by Sickle-Cell Trait against Subtertian Malarial Infection," *British Medical Journal* 1 (1954).

39. Lehmann especially demonstrated this. Concerning the genetic aspects and the mechanism of natural selection in this example, see Theodosius Dobzhansky, *Mankind Evolving: The Evolution of the Human Species* (New Haven, Conn.: Yale University Press, 1962).

40. Cf. Jean Bernard, "Esquisse d'une géographie des maladies du sang," *Annales de géographie* 74 (1965); Jean Bernard and Jacques Ruffié, *Hématologie géographique* (Paris: Masson, 1966).

41. [The concept of biocenosis was coined by German biologist Karl Möbius (1825–1908) in 1877 to designate the interactions between the different species living in a particular ecosystem or community.]

42. Frank W. Preston, "The Commonness, and Rarity, of Species," *Ecology* 29 (1948).

43. [The significance of the mathematical equation given here is difficult to interpret because Grmek fails to indicate what the value of x is. This is even more so given that he applies log series and log normal distribution to disease— i.e., to concepts and not to biological micro-organisms or vectors that can be quantified. However, the equation is clearly inspired by the works of Frank Preston and others working in quantitative ecology. The most important point he derived from the statistical analyses presented in 1969 is that log series and log normal distributions entail that each pathocenosis must have a small number of common diseases and a great number of rare ones. Admitting the difficulties in studying ancient pathocenosis mathematically, Grmek later concluded that it was in fact "impossible to apply true mathematical analysis to the disease of people as ancient as those that are the subject of this book." See Mirko D. Grmek, *Diseases in the Ancient Greek World*, trans. Mireille Muellner and Leonard Mueller (Baltimore, Md.: Johns Hopkins University Press, 1989), 4.]

44. Cf. Carrington B. Williams, *Patterns in the Balance of Nature and Related Problems in Quantitative Ecology* (New York: Academic Press, 1964).

45. This opinion is based on the statistical research of G. Herdan ("The Mathematical Relation between the Number of Diseases and the Number of Patients in a Community," *Journal of the Royal Statistical Society*, Section A 120 [1957]) and on my own preliminary analysis.

46. Richmond C. Holcomb, *Who Gave the World Syphilis? The Haitian Myth* (New York: Froben, 1930).

47. Ellis H. Hudson, *Non-Venereal Syphilis: A Sociological and Medical Study of Bejel* (London: E. S. Livingstone, 1958); "Christopher Columbus and the History of Syphilis," *Acta Tropica* 25 (1968).

48. V. Moller-Christensen, "Evidence of Tuberculosis, Leprosy and Syphilis in Antiquity and the Middle Ages," in *Proceedings of the 19th International Congress of the History of Medicine* (New York: Karger, 1966).

49. This work is led by Jean Meyer and Emmanuel Le Roy Ladurie. See Jean Meyer, "Une enquête de l'Académie de médecine sur les épidémies (1774–1794)," *Annales: Économies, Sociétés, Civilisations* 21 (1966). For the methodological difficulties of this study see Jean-Pierre Peter, "Une enquête de la Société Royale de médecine: malades et maladies à la fin du XVIIIᵉ siècle," *Annales:*

Économies, Sociétés, Civilisations 22 (1967). The author is justified in complaining of "the archaic medical vocabulary, indecipherable at first glance" (718).

50. See Mirko D. Grmek, "Les origines d'une maladie d'autrefois: le scorbut des marins," *Bulletin de l'Institut océanographique de Monaco*, special issue 2 (1968).

51. [Grmek did not write this other paper he announced here, but his book *Diseases in the Ancient Greek World* builds on it.]

2. THE CONCEPT OF EMERGING DISEASE

1. See Mirko D. Grmek, "Problème des maladies nouvelles," in *Sida, épidémies et sociétés* (Lyon, France: Fondation Marcel Mérieux, 1987); "Le sida est-il une maladie nouvelle?" *Médecine et maladies infectieuses* 18 (1988); *History of AIDS: Emergence and Origin of a Modern Pandemic*, trans. Russell C. Maulitz and Jacalyn Duffin (Princeton, N.J.: Princeton University Press, 1990).

2. Charles Nicolle, *Destin des maladies infectieuses* (Paris: Alcan, 1933). See also Aidan Cockburn, *The Evolution and Eradication of Infectious Diseases* (Baltimore, Md.: Johns Hopkins University Press, 1963); F. Henschen, *The History and the Geography of Diseases* (New York: Delacorte, 1966); and Robert Hudson, "How Diseases Birth and Die," *Transactions of the College of Physicians of Philadelphia* 45 (1977).

3. Thierry Bardinet, "Le cœur et le système des 'conduits' (metou) dans les conceptions physiologiques de l'Égypte ancienne" (PhD diss., Université de Paris, EPHE, 1992), 61–62.

4. Thucydides, *History of the Peloponnesian War*, books 1 and 2, trans. Charles Forster Smith (London: William Heinemann, 1956), 343.

5. Thucydides, 343.

6. Thucydides, 343.

7. Pliny the Elder, *The Natural History*, trans. John Bostock (London: Taylor and Francis, 1855); Seneca, *Naturales Quaestiones*, trans. Thomas H. Corcoran (London: Heinmann, 1971). See Mirko D. Grmek, "La dénomination latine des maladies considérées comme nouvelles par les auteurs antiques," in *Le Latin médical (Mémoires du Centre Jean-Palerne, X)*, ed. Guy Sabbah (Saint-Étienne, France: Presses de l'Université de Saint-Étienne, 1991).

8. Plutarch, *Propos de table, Oeuvres morales de Plutarque*, trans. Ricard (Paris: Didier, 1844).

9. See Charles Mugler, "Démocrite et le danger de l'irradiation cosmique," *Revue d'histoire des sciences* 20 (1967).

10. [According to the usually accepted birth/death dates of Galen. See Susan P. Mattern, *The Prince of Medicine: Galen in the Roman Empire* (Oxford: Oxford University Press, 2013), 29–30, 274.]

11. See, for instance, Lloyd G. Stevenson, "New Diseases in the Seventeenth Century," *Bulletin of the History of Medicine* 39 (1965).

12. For more recent developments, see Bernardino Fantini, "La biologica molecolare e il concetto di malattia," in *Maladie et maladies, histoire et conceptualisation (Mélanges en l'honneur de Mirko Grmek)*, ed. Danielle Gourevitch (Geneva: Droz, 1992).

13. Mirko D. Grmek, *Diseases in the Ancient Greek World*, trans. Mireille Muellner and Leonard Muellner (Baltimore, Md.: Johns Hopkins University Press, 1989), 328, 332–33, 335. On the modern emergence of these ancient diseases, see Hans Zinsser, *Rats, Lice and History* (Boston: Little, Brown, 1935), and J. M. Goupil, "L'histoire de la coqueluche" (PhD diss., University of Caen, 1976).

14. Grmek, *Diseases in the Ancient Greek World*, 335–36.

15. On this topic, see the clarifications on recent research concerning the history of epidemic diseases by Erwin H. Ackerknecht, "Causes and Pseudo-causes in the History of Diseases," in *A Celebration of Medical History: The Fiftieth Anniversary of the Johns Hopkins Institute of the History of Medicine and the Welch Medical Library*, ed. Lloyd G. Stevenson (Baltimore, Md.: Johns Hopkins University Press, 1982); Charles E. Rosenberg, "Commentary" on Ackerknecht, in *A Celebration of Medical History*; and the provoking book by Thomas McKeown, *The Origins of Human Diseases* (London: Blackwell, 1988).

16. See, for instance, A. Perrenoud, "Contributions à l'histoire cyclique des maladies. Deux cents ans de variole à Genève (1580–1810)," in *Mensch und Gesundheit in der Geschichte*, ed. Arthur E. Imhof (Husum, Germany: Matthiesen, 1980).

17. William I. Beveridge, *Influenza: The Last Plague, An Unfinished Story of Discovery* (New York: Prodist, 1977); Edward D. Kilbourne, *Influenza* (New York: Plenum, 1987); and Richard E. Shope, "Influenza: History, Epidemiology, and Speculation," *Public Health Reports* 73 (1988).

18. See, for example, Leon Rosen, "Dengue in Greece in 1927 and 1928 and the Pathogenesis of Dengue Hemorrhagic Fever: New Data and a Different Conclusion," *The American Journal of Tropical Medicine* 35 (1986).

19. Henri H. Mollaret, "Interprétation socio-écologique de l'apparition de maladies réellement nouvelles," in *Sida, épidémies et sociétés* (Lyon: Fondation Marcel Mérieux, 1987).

20. Bruce B. Dan, "Toxic Shock Syndrome: Back to the Future," *The Journal of the American Medical Association* 257 (1987).

21. David W. Fraser et al., "Legionnaire's Disease: Description of an Epidemic of Pneumonia," *New England Journal of Medicine* 297 (1977); Joseph E. McDade et al., "Legionnaire's Disease: Isolation of a Bacterium and Demonstration of Its Role in Other Respiratory Diseases," *New England Journal of Medicine* 297 (1977); G. Lattimer and R. A. Osborne, *Legionnaire's Disease* (New York: Dekker, 1981).

22. Grmek, *Diseases in the Ancient Greek World*, 152–76, 198–209. See also Keith Manchester, "Leprosy: The Origin and Development of the Disease in Antiquity," in *Maladie et maladies, histoire et conceptualisation (Mélanges en l'honneur de Mirko Grmek)*, ed. Danielle Gourevitch (Geneva: Droz, 1992). Among recent publications concerning the oldest period, see also Thierry Bardinet, "Remarques sur les maladies de la peau, la lèpre et le châtiment divin dans l'Égypte ancienne," *Revue d'Égyptologie* 39 (1988), and M. Stol, "Leprosy: New Light from Greek and Babylonian Sources," *Jaarbericht van het Vooraziatische-Egyptisch genootshchap. Ex Oriente lux* 30 (1988).

23. Grmek, *Diseases in the Ancient Greek World*, 133–51. See also F. E. Rabello, "Les origines de la syphilis," *Nouvelle presse médicale* 2 (1973); Francisco Guerra, "The Dispute over Syphilis: Europe versus America," *Clio Medica* 13 (1978). Recent paleopathological discoveries bring additional evidence regarding the antiquity of treponematosises. See on this topic B. M. Rothschild and W. Turnbull, "Treponemal Infection in a Pleistocene Bear," *Nature* 329 (1987).

24. Gino Fornaciari et al., "Syphilis in a Renaissance Italian Mummy," *The Lancet* 334 (1989).

25. André Fribourg-Blanc and Henri H. Mollaret, "Natural Treponematosises of the African Primate," *Primates in Medicine* 3 (1968); and André Fribourg-Blanc, G. Niel, and Henri H. Mollaret, "Confirmation sérologique et microscopique de la tréponématose du cynocéphale de Guinée," *Bulletin de la Société de pathologie exotique* 59 (1966).

26. See J. D. Oriel and Aidan Cockburn, "Syphilis: Where Did It Come From?" *Paleopathology Newsletter* 6 (1974); B. J. Baker and G. J. Armelagos, "The Origin and Antiquity of Syphilis," *Current Anthropology* 29 (1988).

27. See Francisco Guerra, "La invasion de America por virus," in *Maladie et maladies, histoire et conceptualisation (Mélanges en l'honneur de Mirko Grmek)*, ed. Danielle Gourevitch (Geneva: Droz, 1992).

28. Esther W. Stearn and Allen E. Stearn, *The Effect of Smallpox on the Destiny of the Amerindians* (Boston: Bruce Humphries, 1945); Percy M. Ashburn, *The Ranks of Death: A Medical History of the Conquest of America* (New York: Coward-McCann, 1947); Alfred W. Crosby, "Conquistador y Pestilencia: The First New World Pandemic and the Fall of the Great Indian Empires," *The Hispanic American Historical Review* 47 (1967).

29. See Francisco Guerra, "Cause of Death of the American Indians," *Nature* 326 (1987); "The Earliest American Epidemic: The Influenza of 1493," *Social Science History* 12 (1988).

30. Ann G. Carmichael and Arthur M. Silverstein, "Smallpox in Europe before the Seventeenth Century: Virulent Killer or Benign Disease?" *The Journal of the History of Medicine and Allied Sciences* 42 (1987).

31. Guerra, "La invasion," 223. See also Henry R. Carter, *Yellow Fever: An Epidemiological and Historical Study of Its Place of Origin* (Baltimore, Md.: Williams and Wilkins, 1931); and Wilbur G. Downs, "History of Epidemiological Aspects of Yellow Fever," *Yale Journal of Biology and Medicine* 55 (1982).

32. See, for instance, H. Harold Scott, *A History of Tropical Medicine* (London: Edward Arnold, 1939); Philip D. Curtin, *Death by Migration: Europe's Encounter with the Tropical World in the Nineteenth Century* (New York: Cambridge University Press, 1989); and Russell C. Maulitz, "Reflections on Yellow Fever and AIDS," in *Maladie et maladies, histoire et conceptualisation (Mélanges en l'honneur de Mirko Grmek)*, ed. Danielle Gourevitch (Geneva: Droz, 1992).

33. The irruption of this disease on the Faroe Islands in 1846 provides a historical example that, in the presentation by Peter Ludwig Panum, takes on the appearance of a genuine epidemiological experience. See Peter Ludwig Panum, "Observations Made during the Epidemic of Measles on the Faroe Islands in the

Year 1846," in *Medical Classics*, vol. 3, ed. Emerson C. Kelly (New York: Williams and Wilkins, 1939).

34. Grmek, *Diseases in the Ancient Greek World*, 177–97; Charles Coury, *Grandeur et déclin d'une maladie. La tuberculose au cours des âges* (Suresnes, France: Lepetit, 1972).

35. M. Monir Madkour, "Historical Aspects of Brucellosis," in *Brucellosis*, ed. M. Monir Madkour (London: Butterworths, 1989).

36. See Frank Macfarlane Burnet, *Viruses and Man* (Harmondsworth, England: Penguin Books, 1955), and Grmek, *Diseases in the Ancient Greek World*, 98–99.

37. See W. Graeme Laver, *The Origin of Pandemic Influenza Viruses* (New York: Elsevier, 1983); see also Beveridge, *Influenza*; Kilbourne, *Influenza*; and Shope, "Influenza."

38. R. Politzer, *La peste* (Geneva: WHO, 1954); Jean-Noël Biraben, *Les hommes et la peste dans les pays européens et méditerranéens* (Paris: Moudon, 1975–1976); William H. McNeill, *Plagues and People* (New York: Doubleday, 1976).

39. René Devignat, "Variétés de l'espèce *Pasteurella pestis*; nouvelle hypothèse," *Bulletin of the World Health Organization* 4 (1951); *La peste antique du Congo belge dans le cadre de l'histoire et de la géographie* (Brussels: Institut Royal Colonial Belge, 1953); and "Répartition géographique des trois variétés de *Pasteurella pestis*," *Schweizerische Zeitschrift für Allgemeine Pathologie* 16 (1953).

40. See especially Henri H. Mollaret, "A Personal View of the History of the Genus *Yersinia*," *Contributions to Microbiology and Immunology* 9 (1987); Richard E. Lenski, "Evolution of Plague Virulence," *Nature* 334 (1988); and Roland Rosqvist, Mikael Skurnik, and Hans Wolf-Watz, "Increased Virulence of *Yersinia* Pseudo-Tuberculosis by Two Independent Mutations," *Nature* 334 (1988).

41. See John D. Frame, John M. Baldwin Jr., David J. Gocke, and Jeanette M. Troup, "Lassa Fever: A New Virus Disease of Man from West Africa," *The American Journal of Tropical Medicine and Hygiene* 19 (1970); Gustav A. Martini and Rudolf Siegert, *Marburg Virus Disease* (Berlin: Springer, 1971); T. P. Monath, "Lassa Fever: Review of its Epidemiology and Epizootiology," *Bulletin of the World Health Organization* 52 (1975); P. Brès, "Les virus Lassa, Marburg et Ebola, nouveaux venus en pathologie tropicale," *Nouvelle presse médicale* 7 (1978); J. W. LeDuc, "Epidemiology of Hemorrhagic Fever Viruses," *Clinical Infectious Diseases* 2, suppl. 4 (1989).

42. R. S. Roberts, "A Consideration of the Nature of the English Sweating Sickness," *Medical History* 9 (1965); J. Brossolet, "Expansion européenne de la suette anglaise," *Proceedings of the XIII International Congress of the History of Medicine, 1972*, vol. 1 (1974): 595–600; John A. H. Wylie and Leslie H. Collier, "The English Sweating Sickness (*Sudor Anglicus*): A Reappraisal," *Journal of the History of Medicine and Allied Sciences* 36 (1981).

43. A. J. Watson, "Origin of *Encephalitis lethargica*," *China Medical Journal* 42 (1928); A. Greenough and J. A. Davis, "*Encephalitis lethargica*: Mystery of the Past or Undiagnosed Disease of the Present?" *The Lancet* 1 (1983).

44. L. A. McNicol and R. N. Doetsch, "A Hypothesis Accounting for the

Origins of Pandemic Cholera: A Retrograde Analysis," *Perspectives in Biology and Medicine* 26 (1983). On the conditions responsible for the pandemic spread, see in particular A. Dodin, "Pourquoi les pandémies cholériques?" *Bulletin de la Société de pathologie exotique* 77 (1984).

45. Mirko D. Grmek, "Preliminaries for a Historical Study of Diseases" [this volume].

46. ["History is life's teacher."]

47. See, in particular, the thoughts of Joshua Lederberg during the conference on emerging viruses held in Washington in 1989 and the articles of Stephen S. Morse and Ann Schluederberg, "Emerging Viruses: The Evolution of Viruses and Viral Diseases," *Journal of Infectious Diseases* 162 (1990); and Stephen S. Morse, "Emerging Viruses: Defining the Rules for Viral Traffic," *Perspectives in Biology and Medicine* 34 (1991).

48. Frank Fenner et al., *The Biology of Animal Viruses* (New York: Academic Press, 1974); B. N. Fields, *Virology* (New York: Raven, 1990).

49. See Øivind Bergh, Knut Yngve Børsheim, Gunnar Bratbak, and Mikal Heldal, "High Abundance of Viruses Found in Aquatic Environments," *Nature* 340 (1989).

3. SOME UNORTHODOX VIEWS AND A SELECTION HYPOTHESIS
ON THE ORIGIN OF THE AIDS VIRUSES

This paper was presented as the Henry E. Sigerist Lecture in the History and Sociology of Medicine, October 7, 1993, at the Yale Medical Historical Library, Connecticut. The author expresses his gratitude to Jacalyn Duffin, Hannah Professor of History of Medicine, Queen's University, Kingston, Ontario, Canada, for her help in the preparation of the English version of this paper.

1. The vague term "new disease" should be replaced by the term "emerging disease," as specified and analyzed in Mirko D. Grmek, "Le concept de maladie émergente," paper read at the International Symposium Emerging Infectious Diseases: Historical Perspectives, Annecy, 1992, and published in *History and Philosophy of the Life Sciences* 15 (1993). On "new" viral diseases, see in particular Stephen S. Morse, ed., *Emerging Viruses* (Oxford: Oxford University Press, 1993).

2. Mirko D. Grmek, *Histoire du sida. Début et origine d'une pandémie actuelle* (Paris: Payot, 1989); translated by Russell C. Maulitz and Jacalyn Duffin as *History of AIDS: Emergence and Origin of a Modern Pandemic* (Princeton, N.J.: Princeton University Press, 1991).

3. For more on the choice of this name by the international commission on virological nomenclature, see Harold E. Varmus, "Naming the AIDS Virus," in *The Meaning of AIDS*, ed. Eric T. Juengst and Barbara A. Koenig (New York: Praeger, 1989).

4. P. H. Duesberg, "Retroviruses as Carcinogens and Pathogens: Expectations and Reality," *Cancer Research* 47 (1987); "Human Immunodeficiency Virus

and Acquired Immunodeficiency Syndrome: Correlation but Not Causation," *Proceedings of the National Academy of Sciences* 86 (1989).

5. For details concerning Duesberg's crusade against the viral theory, see Mirko D. Grmek, "Dark Sides of the Viral Causal Explanation of AIDS," *Croatian Medical Journal* 35 (1994).

6. Essential publications on the distribution of HIV in the body and on the immunopathogenesis of AIDS are quoted in Grmek, "Dark Sides." For an update, see Jean-Claude Ameisen, "Programmed Cell Death and AIDS Pathogenesis," in *Huitième Colloque des Cent-Gardes*, ed. M. Girard and L. Valette (Lyon, France: Mérieux, 1993), and Anthony S. Fauci, "Multifactorial and Multiphasic Components of the Immunopathogenic Mechanisms of HIV Disease," in *Huitième Colloque des Cent-Gardes*, ed. M. Girard and L. Valette, (Lyon, France: Mérieux, 1993).

7. See C. Grunfeld and K. R. Feingold, "Metabolic Disturbances and Wasting in the Acquired Immunodeficiency Syndrome," *New England Journal of Medicine* 327 (1992).

8. J. Leibowitch, "Wasting as the Ultimate Defense, or the Metchnikoff-Gandhi Macrophage," presented at the Seventh Cent Gardes Meeting, Marnes-la-Coquette, France, 1992, but not included in published proceedings.

9. For details, see Grmek, "Dark Sides."

10. On this subject, see especially *The USSR's AIDS Disinformation Campaign* (Washington, DC: Department of State, 1987); Jean-François Revel, *La connaissance inutile* (Paris: Grasset, 1988), 293–97; Grmek, *History of AIDS*, 151–52; and René Sabatier, *Sida: l'épidémie raciste* (Paris: L'Harmattan, 1989), 98–101.

11. V. Zapevalov, "Panic in the West, or What Is behind the Sensation about AIDS," [in Russian], *Literaturnaya Gazeta*, October 30, 1985.

12. B. Bhushan, "AIDS, a Soviet Propaganda Tool," *The Times of India*, November 19, 1986.

13. [In the original paper by Grmek, the wording of the caption was changed. The phrase provided in the text here is as it appears on the cover of the report *Soviet Influence Activities: A Report on Active Measures and Propaganda, 1986–1987* (Washington, DC: United States Department of State, August 1987).]

14. Jakob Segal, Lilli Segal, and Ronald Dehmlow, *AIDS: Its Nature and Origin* (Bertrand Russell Peace Foundation, Australian Branch, 1986).

15. Robert Gallo, *Virus Hunting* (New York: Basic Books, 1991), 220.

16. E.g., articles in *The African Interpreter* (Lagos), October 1986; *Mirror* (Accra), April 1987; *Sunday Times of Kenya* (Nairobi), June 1987; *Le Devoir* (Dakar), July 1987.

17. John Seale, "AIDS Virus Infection: Prognosis and Transmission," *Journal of the Royal Society of Medicine* 78 (1985).

18. *The Sunday Express*, October 26, 1986.

19. See G. Hancock and E. Carim, *AIDS: The Deadly Epidemic* (London: Gollancz, 1986), 72.

20. Zhores A. Medvedev, "AIDS Virus Infection: A Soviet View of Its

Origin," *Journal of the Royal Society of Medicine* 79 (1986) (with the addition of Seale's response).

21. John Seale and Zhores A. Medvedev, "Origin and Transmission of AIDS: Multi-Use Hypodermics and the Threat to the Soviet Union: Discussion Paper," *Journal of the Royal Society of Medicine* 80 (1987); John Seale, "Origins of the AIDS Viruses, HIV–1 and HIV–2: Fact or Fiction? Discussion Paper," *Journal of the Royal Society of Medicine* 81 (1988); and John Seale, "Crossing the Species Barrier: Viruses and the Origin of AIDS in Perspective," *Journal of the Royal Society of Medicine* 82 (1989).

22. Seale and Medvedev, "Origin and Transmission," 302.

23. John Seale, "Artificial HIV?" *Nature* 335 (1988), 391.

24. Seale, "Origins of the AIDS Viruses," 539.

25. Rolande Girard, *Tristes chimères. Sida* (Paris: Grasset, 1987), 80, 96, 112, 124, and 129. See also, Enzo Biagi, *Il sole malato. Viaggio nella paura dell'AIDS* (Milan: Mondadori, 1987), 195–200.

26. Gallo, *Virus Hunting*, 221.

27. Viktor Zhdanov, *Sovteskaya Kultura*, December 5, 1985.

28. F. Nouchi, "La transparence de l'information en URSS doit s'appliquer au sida," *Le Monde*, November 6, 1987.

29. Vincent Jauvert, "La rumeur du KGB," *Le Nouvel Observateur*, June 11, 1992.

30. See Hancock and Carim, *AIDS*, 73.

31. [Grmek did not reference Zhdanov's claim. Also, the report *Soviet Influence Activities* cites Zhdanov's reply slightly differently from Grmek's quote in the original version of his article. The phrase provided in the text, however, is as it appears in *Soviet Influence Activities*, 37.]

32. On Democritus, see Charles Mugler, "Démocrite et le danger de l'irradiation cosmique," *Revue d'histoire des sciences* 20 (1967); for modern hypotheses, see Fred Hoyle and Nalin C. Wickramasinghe, *Diseases from Space* (London: J. M. Dent, 1979), and Francis Crick, *Life Itself: Its Origin and Nature* (New York: Simon and Schuster, 1981).

33. See two excellent review articles: H. Rübsamen-Waigman and U. Dietrich, "Die Ahnen des AIDS-Virus," *Bild der Wissenschaft* 3 (1991), and G. Myers, K. MacInnes, and B. Korber, "The Emergence of Simian/Human Immunodeficiency Viruses," *AIDS Research in Human Retroviruses* 8 (1992).

34. M. Essex and P. J. Kanki, "The Origin of the AIDS Virus," *Scientific American* 259 (1988). The 1988 formulation of the simian origin hypothesis is now outdated. For an update, see M. Essex, "Origin of AIDS," in *AIDS: Etiology, Diagnosis, Treatment and Prevention*, ed. Vincent T. DeVita Jr., Samuel Hellman, and Steven A. Rosenberg, 3rd ed. (Philadelphia, Penn.: Lippincott, 1993).

35. See B. Hahn, "Biologically Unique, SIV-like HIV-a Variants in Healthy West African Individuals," in *Cinquième Colloque des Cents-Gardes*, ed. M. Girard and L. Valette (Lyon, France: Mérieux, 1990).

36. Ursula Dietrich et al., "A Highly Divergent HIV–2 Related Isolate," *Nature* 342 (1989).

37. M. Peeters et al., "Isolation and Partial Characterization of an HIV-Related Virus Occurring Naturally in Chimpanzees in Gabon," *AIDS* 3 (1989); and Thierry Huet et al., "Genetic Organization of a Chimpanzee Lentivirus Related to HIV–1," *Nature* 345 (1990).

38. Robert De Leys et al., "Isolation and Partial Characterization of an Unusual Immunodeficiency Retrovirus from Two Persons of West-Central-Africa Origin," *Journal of Virology* 64 (1990).

39. F. Fontenay and L. De Villepin, "Un entretien avec le Pr. Luc Montagnier," *Journal du Sida* 52–53 (1993). See also L. Montagnier, "Origin and Evolution of HIVs and Their Role in AIDS Pathogenesis," *Journal of Acquired Immunological Deficiency Syndrome* 1 (1988); and "Le rétrovirus de l'immunodéficience chez l'homme et les primates," *Bulletin de l'Académie Nationale de Médecine* 173 (1989).

40. I am preparing a paper on all arguments invoked in favor of or against the cross-transmission of HIV in Africa.

41. For more on the hypothesis that mycoplasmas are cofactors of AIDS, see M. Lemaitre et al., "Protective Activity of Tetracycline Analogs against the Cyctopathic Effect of the Human Immunodeficiency Virus in CEM Cells," *Research in Virology* 141 (1990); M. Lemaitre et al., "Role of Mycoplasma Infection in the Cytopathic Effect Induced by the Immunodeficiency Virus Type 1 in Infected Cell Lines," *Infection and Immunity* 60 (1992); and D. Taylor-Robinson, "Are Mycoplasmas Involved in the Pathogenesis of AIDS?" in *Huitième Colloque des Cents-Gardes*, ed. M. Girard and L. Valette (Lyon, France: Mérieux, 1993).

42. [Here, Grmek is quoting himself. See *History of AIDS*, 155.]

43. [For an empirical study of how the use of inadequately sterilized syringes and needles amplified the spread of HIV in Africa in the early 1980s, against the backdrop of medical campaigns to control various infectious diseases, see Jacques Pépin, *The Origins of AIDS* (Cambridge: Cambridge University Press, 2011). Curiously, Pépin's study does not cite Grmek's own history of AIDS. For a useful comparative analysis of the two books, see J. Duffin, "Best-Intentioned Efforts Aggravated AIDS Pandemic," *Canadian Medical Association Journal* 184 (2012).]

44. See J. Holland et al., "Rapid Evolution of RNA Genomes," *Science* 215 (1982); and S. P. Clark and T. W. Mak, "Fluidity of a Retrovirus Genome," *Journal of Virology* 50 (1984).

45. M. Eigen and P. Schuster, *The Hypercycle: A Principle of Natural Self-Organization* (New York: Springer, 1979). See also N. Eigern, *Die Stufen zum Leben* (Munich: Pipper, 1987). For HIV in particular, see W. A. Haseltine and F. Wong-Staal, "The Molecular Biology of the AIDS Virus," *Scientific American* 259 (1988). [The term "replicator" was made famous by Oxford zoologist Richard Dawkins, who used it in his popular science book *The Selfish Gene* (Oxford: Oxford University Press, 1976).]

46. B. H. Hahn et al., "Genetic Variation in HTVL–III/LAV over Time in Patients with AIDS or at Risk for AIDS," *Science* 232 (1986); Marc Alizon and Luc Montagnier, "Genetic Variability in Human Immunodeficiency Viruses,"

Annals of the New York Academy of Sciences 511 (1987); M. Goodenough et al., "HIV–1 Isolates Are Rapidly Evolving Quasispecies: Evidence for Viral Mixture and Preferred Nucleotide Substitutions," *Journal of Acquired Immune Deficiency Syndromes* 2 (1989); S. Wain-Hobson and J. P. Vartanian, "Sida; suivre la variation du virus," *La Recherche* 23 (1992).

47. M. Nowak, "HIV Mutation Rate," *Nature* 347 (1990).

48. Jennifer Learmont et al., "Long-Term Symptomless HIV–1 Infection in Recipients of Blood Products from a Single Donor," *Lancet* 340 (1992). Cf. also Thierry Huet et al., "A Highly Defective HIV–1 Strain Isolated from a Healthy Gabonese Individual Presenting an Atypical Western Blot," *AIDS* 3 (1989).

49. J. de Mareuil et al., "The Human Immunodeficiency Virus (HIV) Gag Gene Product p18 Is Responsible for Enhanced Fusogenicity and Host Range Tropism of Highly Cytopathic HIV–1–NDK Strain," *Journal of Virology* 66 (1992).

50. [A view traditionally attributed to American comparative pathologist Theobald Smith. See "Preliminaries for a Historical Study of Diseases," this volume.]

51. H. J. Bremermann and Roy M. Anderson, "Mathematical Models of HIV Infection: Threshold Condition for Transmission and Host Survival," *Journal of Acquired Immune Deficiency Syndromes* 3 (1990); Roy M. Anderson, "The Ecological Factors that Determine the Impact of a New Disease in Host Population" (paper presented at the International Symposium on Emerging Infectious Diseases: Historical Perspectives, Mérieux Conference Center, Annecy, France, April 6–8, 1992). [For a critical review of the trade-off model, see Samuel Alizon, A. Hurford, N. Mideo, and M. Van Baalen, "Virulence Evolution and the Trade-Off Hypothesis: History, Current State of Affairs, and the Future," *Journal of Evolutionary Biology* 22 (2009)].

52. Richard E. Lenski, "Evolution of Plague Virulence," *Nature* 334 (1988).

53. Douglas C. Aziz, Zaher Hanna, and Paul Jolicoeur, "Severe Immunodeficiency Disease Induced by a Defective Murine Leukemia Virus," *Nature* 338 (1989).

54. Patricia N. Fultz et al., "Identification and Biologic Characterization of an Acutely Lethal Variant of Simian Immunodeficiency Virus from Sooty Mangabeys (SIV/SMM) AIDS," *AIDS Research and Human Retroviruses* 5 (1989).

55. Louis Pasteur, Charles E. Chamberland, and Émile Roux, "De l'atténuation des virus et de leur retour à la virulence," *Compte Rendus de l'Académie des Sciences* 92 (1881): 435; quoted partially by Anne-Marie Moulin, "La métaphore vaccine. De l'inoculation à la vaccination," *History and Philosophy of the Life Sciences* 14 (1992): 291.

56. G. Cowley, "The Future of AIDS," *Newsweek* 22 (1993). Bernardino Fantini gave an accurate account of my ideas to American researchers during the colloquium AIDS and the Historian, held at Bethesda in 1989 (B. Fantini, "Social and Biological Origins of the AIDS Pandemic," in *AIDS and the His-*

torian, ed. Victoria A. Harden and Guenter B. Risse [Bethesda, Md.: NIH, 1991). Curiously, one distinguished American historian of medicine, Guenter Risse, misunderstood the essential part of my arguments. See Guenter B. Risse, "Review of Mirko D. Grmek. *History of AIDS: Emergence and Origin of a Modern Pandemic*," *Bulletin of the History of Medicine* 65 (1991).

57. Indeed, my book is also cited in this article but only as a sort of pretext for the statement that "in Africa . . . urbanization shattered social structures that had long constrained sexual behavior." Yet well before the publication of this issue, Ruth Marshall, a Paris correspondent of *Newsweek*, had asked me for information about the origin of AIDS and the dangers of emerging viruses. I explained my opinion, but this interview was never printed. It is true that P. Ewald developed and applied to AIDS a thoughtful explanation of the way in which virulence may be related to forms of transmission. Cf. Paul Ewald, "Transmission Modes and the Evolution of Virulence with Special Reference to Cholera, Influenza and AIDS," *Human Nature* 2 (1990). About the related notion of "herd immunity" and other fundamental questions in this new field of evolutionary epidemiology, see also W. O'Connor, "Herd Immunity and the HIV Epidemic," *Preventive Medicine* 20 (1991), and Peter R. Gould, *The Slow Plague: A Geography of the AIDS Pandemic* (Cambridge, Mass.: Blackwell, 1993).

58. See Mirko D. Grmek, "Discussion on Medicine and Culture," in *Medicine and Culture*, ed. Frederick N. L. Poynter (London: Wellcome Institute, 1969), 119; "Preliminaries for a Historical Study of Diseases" [this volume]; and *Diseases in the Ancient Greek World*, trans. Mireille Muellner and Leonard Muellner (Baltimore, Md.: Johns Hopkins University Press, 1989), 2–4.

59. J. Khalife et al., "Immunological Crossreactivity between the Human Immunodeficiency Virus Type 1 Virion Infectivity Factor and a 170-kD Surface Antigen of Schistosoma Mansoni," *Journal of Experimental Medicine* 172 (1990). See also, F. E. G. Cox, "The Worm and the Virus," *Nature* 347 (1990).

60. Edward Tenner, "Revenge Theory," *Harvard Magazine*, March–April (1991).

4. FIRST STEPS IN CLAUDE BERNARD'S DISCOVERY
OF THE GLYCOGENIC FUNCTION OF THE LIVER

1. Mirko D. Grmek, *Catalogue des manuscrits de Claude Bernard. Avec la bibliographie de ses travaux imprimés et des études sur son oeuvre*, foreword by M. Bataillon and E. Wolff, introduction by L. Delhoume and P. Huard (Paris: Collège de France and Masson, 1967).

2. Cf. Claude Bernard, *Philosophie. Manuscrit inédit. Texte présenté par Jacques Chevalier* (Paris: Hatier-Boivin, 1937). See also, Mirko D. Grmek, "Quelques notes intimes de Claude Bernard," *Archives Internationales d'histoire des sciences* 65 (1963). Many important materials are still unpublished. Thus, for example, in the first draft of his acceptance address upon his election to the Académie Française, Bernard expresses some very interesting thoughts that, in fact, were

not intended to be disclosed publicly and that subsequently were omitted from the final lecture.

3. Only a small part of these journals is published. The so-called *Cahier rouge* represents a curious mixture of "philosophical" and technical notes; cf. Claude Bernard, *Cahier de notes 1850–1860. Présenté et commenté par Mirko D. Grmek*, preface by R. Courrier (Paris: Gallimard, 1965). [See also, Claude Bernard, *Claude Bernard and Experimental Medicine: Collected Papers from a Symposium Commemorating the Centenary of the Publication of* An Introduction to the Study of Experimental Medicine *and the First English Translation of Claude Bernard's* Cahier Rouge, ed. Francisco Grande and Maurice B. Visscher (Cambridge, Mass.: Schenkman, 1967).] In one of my recent publications I have used Bernard's manuscript for a detailed analysis of the genesis of an important scientific concept with complicated "philosophical" implications; see Mirko D. Grmek, "Évolution des conceptions de Claude Bernard sur le milieu intérieur," in *Philosophie et méthodologie scientifique de Claude Bernard* (Paris: Masson, 1967).

4. Claude Bernard, *Introduction à l'étude de la médecine expérimentale* (Paris: Baillière, 1865); trans. by H. Copley Greene as *An Introduction to the Study of Experimental Medicine*, with an introduction by L. J. Henderson (New York: Dover, 1949).

5. Cf. Mirko D. Grmek, "Examen critique de la genèse d'une grande découverte: la piqûre diabétique de Claude Bernard," *Clio Medica* 1 (1966).

6. Jean-Baptiste Dumas and Jean Baptiste Boussingault, *Essai de statique chimique des êtres organisés*, 3rd ed. (Paris: Masson, 1844); trans. as *The Chemical and Physiological Balance of Organic Nature: An Essay* (New York: Saxton, 1844).

7. Claude Bernard, "Du suc gastrique et de son rôle dans la nutrition" (thesis for the doctorate in medicine, École de Médecine de Paris, Rignoux, 1843).

8. Bernard, *An Introduction*, 163–64.

9. It is only in his thesis for the doctorate in science that Claude Bernard gives some valuable information concerning the first steps of his discovery of the glycogenic function of the liver. He states that his aim was to follow closely the sugar that was absorbed from food. He wanted to know if it was destroyed in traversing the liver; then what happened after the passage of the bloodstream with sugar through the lungs, and so on. For this purpose, a dog that had been fed on carbohydrate food for seven days was killed during the digestion, and Bernard was able to show that the blood of the hepatic veins, where they join the inferior vena cava, contained a large amount of glucose. This seemed to be an experimental proof that the liver did not destroy the sugar. As a counterproof, Claude Bernard performed a similar experiment on a dog that had been fed exclusively on meat and, to his surprise, found again that the blood of the hepatic veins contained a considerable amount of sugar, although there was no sugar in the intestines. He found also that "the blood of the portal vein contains no sugar before it enters the liver, whereas on leaving that organ the same blood contains considerable amounts of glucose." Cf. Claude Bernard, "Recherches sur une nouvelle fonction du foie considéré comme organe producteur de matière sucrée

chez l'homme et les animaux" (thesis for the degree of doctor *ès-sciences naturelles* [Paris: Martinet, 1853]). In spite of some simplifications and errors (for example, the omission of the fact that in the experiment with the dog on a meat diet, Bernard discovered sugar in the blood of the portal vein), this story is basically correct, especially on its emphasis on Bernard's astonishment after the unexpected results of the counterproof experiment. The best historical presentations in English of Bernard's discovery of the glycogenic function of the liver follow the text of his thesis. For example, D. Wright Wilson, "Claude Bernard," *Popular Scientific Monthly* 84 (1917); F. G. Young, "Claude Bernard and the Theory of the Glycogenic Function of the Liver," *Annals of Science* 2 (1937); and J. M. D. Olmsted, *Claude Bernard, Physiologist* (New York: Harper, 1938). In most other publications, the story is very distorted.

10. Claude Bernard, "De l'origine du sucre dans l'économie animale," *Archives générales de médecine* 18 (1848). An English translation ("The Origin of Sugar in the Animal Body") is published in Emerson Crosby Kelly's *Medical Classics*, vol. 3 (Baltimore, Md.: Williams and Wilkens, 1939).

11. Claude Bernard's unpublished papers in the Collège de France, *Ms. 7b, 7c,* and others. [Bernard's unpublished manuscripts are available online in the Fonds Claude Bernard on the website of the Collège de France at https://salamandre .college-de-france.fr.]

12. He worked in Rayer's and Andral's departments in the well-known hospital, La Charité, in Paris; cf. *Ms. 7b,* 246, 249–50; *Ms. 15i* and *Fasc. 25b,* f. 370.

13. *Ms. 7b,* 133.

14. *Ms. 7b,* 130.

15. [Thomas Willis, in his *Opera Omnia* (Coloniae, 1694), commented on the sweetness of urine in diabetics and wondered "why the urine is wonderfully sweet like sugar or honey."]

16. *Annali universali di medicina e chirurgia* 74 (1835): 160, cited in introductory chapter in R. Lépine, *Le diabète sucré* (Paris: Alcan, 1909).

17. F. Tiedemann and L. Gmelin, *Die Verdauung nach Versuchen*, vol. 1 (Leipzig, Germany: Groos, 1826).

18. Robert MacGregor, "An Experimental Inquiry into the Comparative State of Urea in Healthy and Diseased Urine and the Seat of the Formation of Sugar in Diabetes Mellitus: An Essay," *London Medical Gazette* 20 (1837).

19. Thomas Thomson, *Chemistry of Animal Bodies* (Edinburgh: Adam and Charles Black, 1843).

20. Lépine, *Le diabète sucré.* See also Bernard's historical sketch in *Leçons sur le diabète* (Paris: Baillière, 1877), 142–61.

21. François Magendie, "Note sur la présence normale du sucre dans le sang," *Compte Rendus de l'Académie des Sciences* 23 (1846): 189.

22. In one of his last works, at the very end of his life, Bernard states proudly: "I demonstrated . . . that glycemia is independent from food; that it is found in man and in animals fed with meat or forced to fast. I established that the presence of sugar in blood is a normal fact that always coincides with a state

of health and that disappears only when nutrition is stopped. Thus, instead of admitting with my predecessors that glycemia was a pathological or accidental fact, I have shown the opposite to be true, namely that it is the absence of sugar in blood that is abnormal" (*Leçons sur le diabète*, 127–28 [my translation]).

23. *Ms.* 7c, 308.

24. *Ms.* 7c, 311.

25. He presumed that "an animal in the process of digesting starch should have sugar in the portal arterial vein, not in the venous returning blood" (*Ms.* 7c, 311; note dated May 31, 1848 [my translation]).

26. *Ms.* 7c, 307.

27. *Ms.* 7c, 312.

28. *Ms.* 7c, 354, 358, 363–66.

29. [Grmek does not provide details about the identity of this Irish physician. However, one can find a confirmation of the attribution to Aldridge by Grmek as to the normal presence of sugar in urine by chemist William Sullivan: "I may mention that Aldridge long ago stated that sugar was a normal constituent of the urine." William Sullivan, "On the Progress of General, Physiological, and Pathological Chemistry for the Years 1846–1847," *Dublin Quarterly Journal of Medical Science* 13 (1849): 225.]

30. Claude Bernard and Ch. Barreswil, "Du sucre dans l'œuf," *Compte Rendus de la Société de Biologie* 1 (1849).

31. *Ms.* 7c, 338.

32. [Lymphatic vessels of the small intestine whose function is to collect lipids.]

33. [White liquid contained in the chyliferous vessels.]

34. *Ms.* 7c, 379–82. [This last section of Bernard's quote starting with "This experiment is exceedingly strange" is taken from Holmes's own English translation. Frederic L. Holmes, *Claude Bernard and Animal Chemistry: The Emergence of a Scientist* (Cambridge, Mass.: Harvard University Press, 1974), 423.]

35. [Holmes points out that there was an error in Grmek's original text, giving "reactive" instead of "reagent" as the last word of this sentence. This is now corrected in the present volume. See Holmes, *Claude Bernard and Animal Chemistry*, 529n71.]

36. *Ms.* 7c, 387.

37. *Ms.* 7c, 390.

38. [Person with albumin in the urine; may indicate kidney disease.]

39. *Ms.* 7c, 392.

40. Claude Bernard and Ch. Barreswil, "De la présence du sucre dans le foie," *Compte Rendus de l'Académie des Sciences* 27 (1848).

41. Claude Bernard, "De l'origine du sucre dans l'économie animale," *Archives générales de médecine* 4 (1848).

42. James M. D. Olmsted and E. Harris Olmsted, *Claude Bernard and the Experimental Method in Medicine* (New York: Schuman, 1952), 101.

43. Cannon's book offers many excellent examples of "deductive," histori-

cally wrong approaches to the analysis of scientific discoveries. Cannon states of Claude Bernard:

> In testing the blood for its sugar content at various points after its departure from the intestine, where sugar is absorbed, he found less in the blood of the left side of the heart and in the arteries than in the veins. He drew the erroneous conclusion that the sugar was consumed in the lungs. Then Bernard's interest in the metabolism of sugar in the body led him to examine persons suffering from diabetes, and he was struck by the evidence that the output of sugar in the urine of diabetics is greater than that represented in the food they take in. There sprang into his mind a guiding idea that sugar is *produced* in the organism.

This is, I guess, the way in which Cannon would discover animal glycogenesis, but it has no connection with historical reality. Walter B. Cannon, *The Way of an Investigator: A Scientist's Experiences in Medical Research* (New York: Norton, 1945), 65.

5. THE CAUSES AND THE NATURE OF AGEING

1. This monograph is mainly based on the author's paper in Croat in "Sympozion o gerontologiji" [The symposium on gerontology] published by the Yugoslav Academy of Sciences and Arts, Zagreb, 1958. The introductory essay is based on the ideas put forward in a short article: Mirko D. Grmek, "Les aspects historiques des problèmes fondamentaux de la gérontologie," *Le Scalpel* 110 (1957).

2. Ignatz L. Nascher, "Geriatrics," *New York Medical Journal* 90 (1909).

3. Alphonse de Lamartine, "Le lac de b . . . ," in *Méditations poétiques* (Paris: Librarie grecque-latine-allemande, 1820).

4. [These first paragraphs presenting an overview of Grmek's theory of ageing are taken from the opening essay of the collection he edited. Mirko D. Grmek, "Definitions," in Mirko D. Grmek, *On Ageing and Old Age: Basic Problems and Historic Aspects of Gerontology and Geriatrics* (The Hague: W. Junk, 1958), 2–3.]

5. Old and new hypotheses concerning the causes of ageing are presented in the following works: Luigi Luciani, *Human Physiology* (London: Macmillan, 1921); G. Stanley Hall, *Senescence: The Last Half of Life* (New York: Appleton, 1922); Emil Abderhalden, "Wandlungen in der Auffassung des Wesens der Alternsvorgänge," *Bulletin der Schweizerischen Akademie der Medizinischen Wissenschaften* 6 (1950); F. Henschen, "La nature réelle du vieillissement," *Revue médicale de Bruxelles* 33 (1953); D. Kotovsky, "Alte une neue Wege in der Erfoschung des Alterns," *Sudhoffs Archiv fur Geschichte der Medizin und der Naturwissenschaften* 38 (1954); M. Bürger, *Altern und Krankheit*, 3rd ed. (Leipzig, Germany: Georg-Thieme, 1957). Historical reviews may also be found in the works quoted below as the introductory part of the author's own hypotheses.

6. Aristotle, *On Youth and Old Age, Life and Death, and Respiration*, trans. W. Ogle (New York: Longmans, Green, 1902). Cf. Max Neuberger, *Geschichte der Medizin*, vol. 1 (Stuttgart, Germany: Verlag, 1906); D'Arcy W. Thompson, *On Aristotle as a Biologist, with a Prooemion on Herbert Spencer* (Oxford: Clarendon, 1913); J. J. Griffin, "Aristotle's Observations on Gerontology," *Geriatrics* 5 (1950).

7. Luciani, *Human Physiology*, 293.

8. ["What, after all, is old age other than the pathway to death? And since death is the extinguishing of the innate heat, old age is, as it were, its fading away." *Galen: Selected Works*, trans. Peter. N. Singer (Oxford: Oxford University Press, 1997), 236.]

9. Galen, *Adversus Lycum Libellus*, in *Galeni adversus Lycum et adversus Iulianum libelli*, Corpus mediocorum Graecorum, ed. E. Wenkebach, vol. 7 (Berlin: Verlag, 1951). Cf. Galen, *Claudii Galeni, Opera omnia* ed. Carl G. Kühn (Leipzig, Germany: Cnobloch, 1921), xviii A2 238.

10. André Du Laurens, *Discours de la conservation de la vue, des maladies mélancoliques, des catarrhes, et de la vieillesse* (Paris: J. Mettayer, 1597).

11. André Du Laurens, "Quatrième discours auquel est traitté de la vieillesse, et comme il la faut entretenir," in *Les Oeuvres de M. André du Laurens* (Rouen, France: Raphael du petit val, 1621), 50.

12. Aurelio Anselmi, *Gerocomica, sive, De senum regimine: opus non modo philosophis & medis gratum, sed hominibus utile* (Venice: Apud Franciscum Ciottum, 1606).

13. "Senectus . . . esse viventis defectum ex debilitate calidi innati ob naturelm diminutionem humidis radicalis." Anselmi, *Gerocomica*. [This citation in Latin can be found in L. Luciani, *Human Physiology*, 294.]

14. Gabriele Zerbi, *Gerontocomia, scilicet de senium cura atque victu* (Rome: Prologus, 1489). Cf. L. Münster, "Il primo trattato pratico compiuto sui problem della vecchiaia," *Rivista di Gerontologia e Geriatria* 1 (1951).

15. Chrisoph W. Hufeland, *Die Kunst das menschliche Leben zu verlängern* (Jena, Germany: Akademische Buchhandlung, 1797).

16. Max Rubner, *Das Problem der Lebensdauer une seine Beziehungen zu Wachstum und Ernährung* (Berlin: Oldenbourg, 1908).

17. Otto Bütschli, "Gedanken über Leben und Tod," *Zoologischer Anzeiger* 5 (1882).

18. Jacques Loeb, "Ueber die Ursache des natürlichen Todes," *Pflügers Archiv für die gesamte Physiologie* 124 (1908); "Natural Death and the Duration of Life," *Scientific Monthly* 9 (1919).

19. Cf. D. Kotovsky, "Alte und neue."

20. Élie Metchnikoff, "Études biologiques sur la vieillesse," *Annales de l'Institut Pasteur* 16 (1901–1902); *Études sur la nature humaine: Essai de philosophie optimiste* (Paris: Masson, 1903); *The Prolongation of Life: Optimistic Studies* (London: Heinmann, 1907); Heinz Zeiss, *Elias Metchnikoff, Leben und Werk* (Jena, Germany: G. Fischer, 1932), and G. P. Saharov, *La Lutte contre la vieillesse selon Metchnikoff* (Moscow, 1938).

21. Alexis Carrel, "Rejuvenation of Culture Tissues," *Journal of American Association of Medicine* 57 (1911); Alexis Carrel and A. H. Ebeling, "Antagonistic Growth Principles of Serum and Their Relation to Old Age," *Journal of Experimental Medicine* 38 (1923); Alexis Carrel and Charles A. Lindbergh, *The Culture of Organs* (New York: Hoeber, 1938).

22. Alexis Carrel, *Man: The Unknown* (New York: Harper and Brothers, 1935).

23. Santorio Santorio, *Medicina Statica: Being the Aphorisms of Sanctorius*, trans. John Quincy (London: W. and Newton in Little Britain, A. Bell at the Cross-Keys in Cornhill, W. Taylor at the Ship in Paternoster-Row, and J. Osborn at the Oxford-Arms in Lombard-Street, 1720). Cf. Mirko D. Grmek, *Santorio Santorio i njegovi aparati i instrumenti* [Santorio Santorio and his apparatus and instruments] (Zagreb: Jugoslavenska akademija znanosti i umjetnosti, 1952).

24. Santorio, *Medicina Statica*, 92.

25. Santorio, 94.

26. Francis Bacon, *Historia vitae et mortis* (London: Lownes, 1623), 9; Francis Bacon, *Sylva Sylvarum, or A Natural History in Ten Centuries. Whereunto Is Newly Added the History Natural and Experimental of Life and Death, or of the Prolongation of Life* (London: J. R., 1650). Cf. F. D. Zeman, "Life's Later Years: Studies in the Medical History of Old Age," *Journal of Mount Sinai Hospital* 12 (1945–1946).

27. Alexis Carrel, "Physiological Time," *Science* 74 (1931); Pierre Lecomte du Noüy, *Le temps et la vie* (Paris: Gallimard, 1936). Cf. H. Benjamin, "Biologic versus Chronologic Age," *Journal of Gerontology* 2 (1947).

28. C. S. Minot, "On the Nature and Cause of Old Age," *Harvey Lecture* 1 (1906); *The Problem of Age, Growth and Death: A Study on Cytomorphosis* (London: Murray, 1908).

29. Robert Rössle, *Wachstum und Altern* (Munich: J. F. Bergmann, 1923). Cf. D. Kotovsky, "R. Rössle und die Altersforschung," *München Medical Wochenschr* 99 (1957).

30. August Weismann, *Ueber die Dauer des Lebens; ein Vortrag* (Jena, Germany: G. Fischer, 1882); *Ueber Leben und Tod* (Jena, Germany: G. Fischer, 1884).

31. Du Laurens, *Discours de la conservation de la vue*.

32. M. S. Mühlmann, *Ueber die Ursache des Alterns* (Wiesbaden, Germany: Bergmann, 1900); "L'état actuel de la question du vieillissement," *Scientia, Milano* 60 (1936); *Uchenie o roste, starosti I smerti* [A study of growth, ageing, and death] (Baku, 1926).

33. Cf. C. and O. Vogt, "Ageing of Nerve Cells," *Nature* 158 (1946); F. M. Townsend, "Changes in Brain with Age," *Journal of Gerontology* 1 (1946); P. Divry, "Considération sur le vieillissement cérébral," *Journal belge de neurologie et de psychiatrie* 47 (1947); L. Binet and F. Bourlière, *Précis de gérontologie* (Paris: Masson, 1955); Bürger, *Altern und Krankheit*.

34. Ivan P. Pavlov, *Polnoe sobranie sočinenij* [Complete works], 2nd ed. (Moscow: Akademii Nauk SSSR, 1951–1952); *Sämtliche Werke* [Complete works] (Berlin: Akademie, 1954–1955). Cf. F. M. Thomas, "Pavlov's Work on Higher

Nervous Activity and Its Development in the USSR," *Nature* 154 (1944); Zhores A. Medvedev, "Rol nervnoi sistemy v procese starenija organizma," [The role of the nervous system in the process of ageing] *Prioda* 3 (1953).

35. Rössle, *Wachstum und Altern*. Cf. Kotovsky, "R. Rössle und die Altersforschung."

36. Alexander A. Bogomolets, *Prodlenie zhini* (Kiev, 1938); *The Prolongation of Life*, trans. P. V. Karpovich and S. Bleeker (New York: Sloan and Pearce, 1946). Cf. D. A. Halpern, "Alexander A. Bogomolets," *American Review of Soviet Medicine* 1 (1943).

37. Charles E. Brown-Séquard, "Des effets produits chez l'homme par des injections sous-cutanées d'un liquide retiré des testicules frais de cobayes et de chiens," *Compte rendus de la Société de Biologie* 41 (1889).

38. [This is chapter 5 ("Rejuvenation") in Grmek's original publication.]

39. Arnold Lorand, "Quelques considérations sur les causes de la sénilité," *Compte rendus de la Société de Biologie* 57 (1904); "Problem of Rejuvenation," *Lancet* 1 (1931); *Das Altern, seine Ursache und Behandlung* (Leipzig, Germany: J. A. Barth, 1932).

40. Constantin I. Parhon, *Bătrânet ea s i tratamentul ei problema reîntineririi* [Old age and its treatment] (Bucharest: Editura Academiei Republicii Populare Române, 1948); *Biologia vîrstelor. Cercetari clinice si exeperimentale* [The biology of ageing: Experimental and clinical research] (Bucharest: Editura Academiei Republicii Populare Române, 1955).

41. Cf. M. A. Goldzreher, "Endocrine Aspects of Senescence," *Geriatrics* 1 (1946).

42. Rubner, *Das Problem der Lebensdauer*.

43. G. Marinescu, "Mécanisme chimico-colloidal de la sénilité et le problème de la mort naturelle," *Revue des sciences* 1 (1914); "Nouvelle contribution à l'étude du mécanisme de la vieillesse," *Bulletin de l'Académie de Médecine* 111 (1934).

44. Auguste Lumière, *Théorie colloidale de la biologie et de la pathologie* (Paris: Chiron, 1922); *Sénilité et rajeunissement* (Paris: André Lesot, 1932).

45. Vladislav Růžička, "Beitrage zum Studium der Protoplasmahysteresis und der hysteretischen Vorgänge, I. Die Protoplasmahysteresis als Entropieerscheinung," *Archiv für Mikroskopiche Anatomie Entwicklung Mechanik* 101 (1924).

46. T. B. Robertson, *The Chemical Basis of Growth and Senescence* (Philadelphia, Penn.: Lippincott, 1923); D. Reichinstein, *Das Problem des Alterns und die Chemie der Lebensvorgänge*, 2nd ed. (Zurich: Akerets Erben, 1940); A. de Gregorio Rocasolano, "Physikalisch-chemische Hypothese über das Altern," *Kolloidchemie Beihefte* 19 (1924); N. R. Dhar, "Old Age and Death from a Chemical Point of View," *Journal of Physical Chemistry* 30 (1926); H. Lampert, "Die kolloidchemische Seite des Alterns und ihre Bedeutung für die Entstehung und Behandlung einiger Krankheiten," *Zeitschfrift Alternsforschung* 1 (1938).

47. A. V. Nagornyi, *Problema starenija i dolgoletija* [Problems of ageing and longevity] (Kharkiv, Ukraine: Kharkov State University Press,1940); *Starenije i prodlenije zhizni* [Ageing and the prolongation of life] (Moscow, 1950).

Nagornyi's theoretical deductions are criticized by Zhores A. Medvedev, "Teorija prof. A. V. Nagornogo o stareniji organizma," *Fiziolgicheskii zhurnal SSSR* 38 (1952).

48. Andrej O. Župančič, *Uvod u opstu patofiziologiju coveka* [Introduction to the general pathophysiology of man] (Belgrade, 1952).

49. H. Kunze, *Forschung und Fortschritte* 9 (1925): 25. Cf. Bürger, *Altern und Krankheit.*

50. A. Pütter, "Lebensdauer und Alternsfaktor," *Zeitschrift für allgemeine Physiologie* 19 (1921). Cf. K. Miescher, "Zur Frage der Alternsforschung," *Experientia* 11 (1955); and G. Schlomka, "Über Ziele und Wege klinischer Alternsforschung," in *Festschfift M. Bürger* (Leipzig, Germany: Wiss Z., 1955).

51. Růžička, "Beitrage zum Studium."

52. Alexis Comfort, *The Biology of Senescence* (New York: Rinehart, 1956).

53. W. Kuhn, "Optische Aktivität und Begrenztheit der Lebensdauer," *Z. Altersforsch* 1 (1939); "Mögliche Beziehungen der optischen Aktivität zum Problem des Alterns," *Experientia* 11 (1955).

6. A SURVEY OF THE MECHANICAL INTERPRETATIONS OF LIFE
FROM THE GREEK ATOMISTS TO THE FOLLOWERS OF DESCARTES

1. Plato, *Timaeus*, trans. Benjamin Jowett (New York: Macmillan, 1987), 74a–b.

2. [The use of mechanical models such as catapults to explain animal motion is not in the cited passage of *On the Motion of Animals*. Aristotle's reliance on such artifacts to characterize living motion, however, was studied by Alfred Espinas in a text cited by Georges Canguilhem in *Knowledge of Life* to which Grmek refers to later on in the present essay. See Alfred Espinas, "L'organisme ou la machine vivante en Grèce au IVᵉ siècle avant J. C.," *Revue de métaphysique et de morale* 11 (1903). As to Canguilhem, he claimed that "it is indisputable that Aristotle found in the construction of war machines such as catapults license to liken the movements of animals to mechanical, automatic movements." See Georges Canguilhem, "Machine and Organism," in *Knowledge of Life*, trans. Stefanos Geroulanos and Daniela Ginsburg (New York: Fordham University Press, 2008), 79.]

3. Aristotle, *Politics*, trans. H. Rackman (Cambridge, Mass.: Harvard University Press, 1932), 1253b23–54b20.

4. Galen, *On the Usefulness of the Parts of the Body*, trans. Margaret Tallmadge May (Ithaca, N.Y.: Cornell University Press, 1968).

5. Andreas Vesalius, *On the Fabric of the Human Body*, trans. W. F. Richardson, in collaboration with J. B. Carman (San Francisco: Norman, 1998).

6. Gómez Pereira. *Antoniana Margarita: opus nempe physicis, medicis ac theologis non minus vtile quam necessarium. Per Gometium Pereiram, medicum Methinæ Duelli, quae Hispanorum lingua Medina de el Campo apellatur, nunc primum in lucem æditum* (Medina del Campo, Spain, 1554), 7.

7. Pereira, 17.

8. Pereira, 17.

9. Ivan P. Pavlov, *Conditioned Reflexes and Psychiatry: Lectures on Conditioned Reflexes*, vols. 1–2, trans. W. H. Gantt (New York: International Publishers, 1927).

10. Pereira, *Antoniana Margarita*, 17.

11. Pereira, 17.

12. [In support of his claim, Grmek gave the following citation from Pereira, which could not be found in his book: "*In hominibus ac brutis nutricatione similis motus his quarti generis conspici.*"]

13. ["Tropism theory" is usually attributed to American biologist Jacques Loeb (1859–1924), who put forth a mechanical view of living organisms' motion.]

14. [In this passage, Grmek is drawing on and paraphrasing the work of Robert Southey (1774–1843), who mentions Pereira and the attack by Palacios, as well as the possibility that Descartes adopted the concept of beast-machine from the Spanish physician. See Robert Southey, *Letters Written during a Short Residence in Spain and Portugal*, 2nd ed. (Bristol: Bulgin and Rosen, 1799), 98–99.]

15. William Harvey, *An Anatomical Disquisition on the Motion of the Heart and Blood in Living Beings*, trans. Robert Willis (London: J. M. Dent, 1906).

16. Santorio Santorio, *Medicina Statica: Being the Aphorisms of Sanctorius*, trans. John Quincy (London: W. and Newton in Little Britain, A. Bell at the Cross-Keys in Cornhill, W. Taylor at the Ship in Paternoster-Row, and J. Osborn at the Oxford-Arms in Lombard-Street, 1728).

17. René Descartes, *Discourse on the Method of Rightly Conducting the Reason and Seeking the Truth in the Sciences*, in *The Philosophical Writings of Descartes*, trans. John Cottingham, Robert Stoothoff, and Dugald Murdoch (Cambridge: Cambridge University Press, 1985), 1:119.

18. René Descartes, *Treatise on Man*, in *The Philosophical Writings of Descartes*, trans. John Cottingham, Robert Stoothoff, and Dugald Murdoch (Cambridge: Cambridge University Press, 1985).

19. René Descartes, "Letter to Henry More, February 5, 1649," in *The Philosophical Writings of Descartes: The Correspondence*, trans. John Cottingham, Robert Stoothoff, Dugald Murdoch, and Anthony Kenny (Cambridge: Cambridge University Press, 1991), 3:366.

20. Descartes, "Letter to Henry More," 3:366.

21. Étienne Gilson, "Descartes, Harvey et la scolastique," in *Études de philosophie médiévale* (Strasbourg, France: Palais de l'Université, 1921).

22. [This point was made earlier by Canguilhem: "It is necessary, in order to understand the animal-machine, to see it as having been preceded, logically and chronologically, both by God as efficient cause and by a pre-existing living being as formal and final cause to be imitated," in "Machine and Organism," 85.]

23. Descartes, *Discourse on the Method*, 1:141.

24. Mirko D. Grmek, "La notion de fibre vivante chez les médecins de l'école iatrophysique," *Clio Médica* 5 (1970).

25. [The expression "animated anatomy" Grmek uses probably comes from Henry Sigerist: "Harvey was an anatomist, and he gave anatomy a new structure

—he made it into *anatomia animata*, or physiology." Henry E. Sigerist, *Man and Medicine: An Introduction to Medical Knowledge*, trans. Margaret Galt Boise, with an introduction by William H. Welch (New York: Norton, 1932), 30. Another passage from Sigerist's book (in French translation), with the same expression, is also cited by Canguilhem. See Georges Canguilhem, *The Normal and the Pathological*, trans. Carolyn R. Fawcett, in collaboration with Robert S. Cohen, introduction by Michel Foucault (New York: Zone Books, 1991), 205].

26. Albrecht von Haller, *Physiological Elements of the Human Body* (1757–1766).

27. Giovanni Alfonso Borelli, *De motu animalium* (Rome: Angeli Bernabò, 1680–1681).

28. Charles Darwin, *On the Origins of Species by Means of Natural Selection, or the Preservation of Favoured Races in the Struggle for Life* (London: Murray, 1859).

29. John Ray, *The Wisdom of God Manifested in Works of Creation* (London: R. Harbin, 1691).

30. Niels Stensen, "Discourse of the Anatomy of the Brain," in *Nicolas Steno: Biography and Original Papers of a 17th Century Scientist*, ed. Troels Kardel and Paul Maquet (Heidelberg, Germany: Springer, 2013), 127–28.

31. Giorgio Baglivi, "De praxi medica," in *Opera omnia medico-pratica et anatomica* (Venice, 1727), 78.

32. Canguilhem, "Machine and Organism," 78.

33. [See Grmek, "La notion de fibre vivante."]

34. Paul-Joseph Barthez, *Nouvelle méchanique des mouvements de l'homme et des animaux* (Paris: Méquignon l'aîné, an VI, 1798).

35. Friedrich Hoffmann, *La médecine raisonnée*, 9 vols., trans. Jacques-Jean Bruhier (Paris: Briasson, 1739–1743).

36. Descartes, *Discourse on the Method*, 1:140.

37. Julien Offroy de la Mettrie, *Man a Machine*, trans. Gertrude C. Bussey, revised by M. W. Calkins (Chicago: Open Court, 1912).

38. Aram Vartanian, *La Mettrie's "L'Homme Machine": A Study in the Origins of an Idea* (Princeton, N.J.: Princeton University Press, 1960), 18.

39. La Mettrie, *Man a Machine*, 93.

40. La Mettrie, 89

41. La Mettrie, 99.

42. Frederick the Great, "Eulogy on Julien Offroy de La Mettrie," in La Mettrie, *Man a Machine*, 6.

43. André Doyon and Lucien Liaigre, *Jacques Vaucasson, mécanicien de génie* (Paris: Presses Universitaires de France, 1966).

44. Excerpts from the minutes of the two conferences by Vaucanson on August 9 and October 6 at the Académie des Sciences et beaux-arts of Lyon in 1741, cited in André Doyon and Lucien Liaigre, "Méthodologie comparée du biomécanisme et de la mécanique comparée," *Dialectica* 10 (1956): 298.

45. Salomon Reisel, "Statua humana circulatoria," in *Miscellana Curiosa, Decuria I, Annus IX/X für 1678/9, Observatio I* (Stuttgart, Germany: Sipho Würrtembergicus, 1690).

46. Claude-Nicolas Le Cat, *Dissertation sur un homme artificiel dans lequel on verrait plusieurs phénomènes de l'homme vivant* (Académie de Rouen, 1744).

47. La Mettrie, *Man a Machine*, 90.

48. Joseph Needham, *Man a Machine* (London: Kegan Paul, 1927), 7.

49. [This citation attributed to Huxley by Grmek was not found. It could be a reference to *The Physics and Chemistry of Life*, ed. Gerard Piel et al. (New York: Simon and Schuster, 1955), or to H. G. Wells, Julian Huxley, George Philip Wells, *The Science of Life*, and especially Book One ("The Body Is a Machine") (New York: The Literary Guild, 1929.]

50. Needham, *Man a Machine*, 92–93.

7. A PLEA FOR FREEING THE HISTORY OF SCIENTIFIC DISCOVERIES FROM MYTH

1. Stendhal, *On Love*, trans. Philip Sidney Woolf and Cecil N. Sidney Woolf (New York: Brentano, 1915), 24.

2. [Leopold von Ranke (1795–1886), often hailed as the founder of "scientific history." Ranke's "precept" usually refers to the idea of writing history "as it really was."]

3. Israel Scheffler, *The Anatomy of Inquiry: Philosophical Studies in the Theory of Science* (New York: Knopf, 1963).

4. Mirko D. Grmek, "Définition du domaine propre de l'histoire des sciences et considérations sur ses rapports avec la philosophie des sciences," *History and Philosophy of the Life Sciences* 1 (1979).

5. L. Pearce Williams, "Normal Science, Scientific Revolutions and the History of Science," in *Criticism and the Growth of Knowledge*, ed. I. Lakatos and A. Musgrave (Cambridge: Cambridge University Press, 1970), 50.

6. Walter Pagel, *New Light on William Harvey* (Basel, Switzerland: Karger, 1976); Howard E. Gruber, *Darwin on Man: A Psychological Study of Scientific Creativity* (New York: Dutton, 1974); Frederic L. Holmes, *Claude Bernard and Animal Chemistry: The Emergence of a Scientist* (Cambridge, Mass.: Harvard University Press, 1974); Mirko D. Grmek, *Raisonnement expérimental et recherches toxicologiques chez Claude Bernard* (Geneva: Droz, 1973).

7. See Grmek, *Raisonnement expérimental*. [For increased clarity, I translate Grmek's expression as the "lived experience of discovery."]

8. [The letters were added to increase the clarity of these passages.]

9. James D. Watson, *The Double Helix: A Personal Account of the Discovery of the Structure of DNA* (London: Weidenfeld and Nicholson, 1968); Robert Olby, *The Path to the Double Helix* (London: Macmillan, 1974).

10. Mirko D. Grmek, "Examen critique de la genèse d'une grande découverte: la piqûre diabétique de Claude Bernard," *Clio Medica* 1 (1966).

11. Ludovico Geymonat, *Galileo Galilei*, 2nd ed. (Turin, Italy: Einaudi, 1962).

12. ["Vagusstoff" is a substance released by the stimulation of the cranial nerve called the vagus nerve and causes a reduction in heart rate.]

13. Zenon M. Bacq, *Les transmissions chimiques de l'influx nerveux* (Paris: Gauthier-Villars, 1974).

14. Abel Rey, *L'énergétique et le mécanisme. Au point de vue des conditions de la connaissance* (Paris: Félix Alcan, 1908).

15. [This chapter was presented as a conference paper during an international school in the history of science in 1977 where Grmek led several discussions, hence the frequent mention of other "seminars he has given."]

16. [Grmek also makes this point in "Définition du domaine," 9.]

17. ["On the way to truth."]

18. Grmek, *Raisonnement expérimental*, 161–69.

19. Gaston Bachelard, *The Formation of the Scientific Mind: A Contribution to a Psychoanalysis of Objective Knowledge*, trans. Mary McAllester Jones (Manchester, UK: Clinamen, 2002).

20. Thomas S. Kuhn, *The Structure of Scientific Revolutions* (Chicago: University of Chicago Press, 1962).

21. Grmek, *Raisonnement expérimental*.

22. [Grmek does not name this scholar.]

23. Catherine de Buzon [now Chevalley], "Remarques sur l'interprétation de l'œuvre de Kepler," *Archives Internationales d'histoire des sciences* 27 (1977): 75.

24. William Harvey, "An Anatomical Disquisition on the Motion of the Heart and Blood in Living Beings" in *The Works of William Harvey*, trans. Robert Willis (Philadelphia: University of Pennsylvania Press, 1989), 3.

25. Mirko D. Grmek, "Le rôle du hasard dans la genèse des découvertes scientifiques," *Medicina nei Secoli* 13 (1976). [With the exception of the last sentence, this whole paragraph was omitted in the originally published version and has now been restored thanks to the original French manuscript retrieved in the Fonds Mirko Grmek. See Mirko D. Grmek, "Plaidoyer pour la démythification des présentations historiques des découvertes scientifiques," 1977, GRM 19, p. 43, Grmek Papers, IMEC, Caen, France.]

26. Peter Razzell, *Edward Jenner's Cowpox Vaccine: The History of a Medical Myth* (Lewes, Del.: Caliban, 1977).

27. Ian Murray, "Paulesco and the Isolation of Insulin," *Journal of the History of Medicine and Allied Sciences* 26 (1971).

28. Joseph H. Pratt, "A Reappraisal of Researches Leading to the Discovery of Insulin," *Journal of the History of Medicine and Allied Sciences* 9 (1954).

29. Massimo Baldini, *Epistemologia e storia della scienza* (Florence: Città di Vita, 1974).

30. Norwood Russel Hanson, "The Idea of a Logic of Discovery," *Dialogue: Canadian Philosophical Review* 4 (1965); Kuhn, *Structure of Scientific Revolutions*.

31. Paul Feyerabend, *Against Method: Outline of an Anarchist Theory of Knowledge* (London: New Left Books, 1975).

32. Jean Piaget, *Biology and Knowledge: An Essay on the Relations between Organic Regulations and Cognitive Processes* (Chicago: University of Chicago Press, 1971).

33. René Thom, *Modèles mathématiques de la morphogenèse. Recueil de textes sur la théorie des catastrophes et ses applications* (Paris: Union générale d'éditions, 1974).

34. Claude Bernard, *An Introduction to the Study of Experimental Medicine*, trans. H. Copley Greene (New York: Schuman, 1949), 24.

35. [Bernard's exact words are: "In sciences of experimentation, man observes, but in addition he acts on matter, analyzes its properties and to his own advantage brings about the appearance of phenomena which doubtless always occur according to natural laws, but in conditions which nature often has not yet achieved. With the help of these active experimental sciences, man becomes an inventor of phenomena, a real foreman of creation." Bernard, *An Introduction to the Study of Experimental Medicine*, 18.]

36. William Stanley Jevons, *The Principles of Science: A Treatise on Logic and Scientific Method* (London: Macmillan, 1874).

37. Émile Meyerson, *Du cheminement de la pensée* (Paris: Alcan, 1931).

38. ["Every theory comes from another theory."] Mirko D. Grmek, "La notion de fibre vivante chez les médecins de l'école iatrophysique," *Clio Medica* 5 (1970).

39. [Here, I translate the French *"gnoséologique"* as "epistemological."]

40. Norwood Russel Hanson, *Patterns of Discovery* (London: Cambridge University Press, 1958).

41. Nicolaas A. Rupke, *"Bathybius Hackelii* and the Psychology of Scientific Discovery," *Studies in History and Philosophy of Science Part A* 7 (1976).

42. Ronald A. Fisher, "Has Mendel's Work Been Rediscovered?" *Annals of Science* 1 (1936).

43. B. L. van der Waerden, "Mendel's Experiments," *Centaurus* 12 (1968); Franz Weiling, "Neue ergebnisse zur statistischen Vorgeschichte der Mendelschen Versuche," *Biometrics* 27 (1971).

44. For a good exposition on the exegesis of cell theory, see Florkin's, Klein's, and Canguilhem's publications. Marcel Florkin, *Naissance et déviation de la théorie cellulaire dans l'œuvre de Th. Schwann* (Paris: Hermann, 1960); Marc Klein, *Histoire des origines de la théorie cellulaire* (Paris: Hermann, 1936); Georges Canguilhem, *Knowledge of Life*, trans. Stefanos Geroulanos and Daniela Ginsburg (New York: Fordham University Press, 2008).

45. Bachelard, *The Formation of the Scientific Mind*.

46. [The adjective "périmée" is a reference to the epistemological work of Bachelard.]

47. Dario Antiseri, "Prova di une teoria ed educazione al riconosciemento dell 'errore,'" *Medicina nei Secoli* 13 (1976).

48. [In this passage Grmek is possibly referring to the work of Popper's disciple, the Hungarian philosopher and logician Imre Lakatos (1922–1974). However, Grmek's reference list does not mention Lakatos's published work "The Methodology of Scientific Research Programs" but rather the book Lakatos never finished because of his death in 1974: *The Changing Logic of Scientific Discovery*.]

49. [The French manuscript gives in parenthesis "(Carnap, Nagel)." However, only the second is included in the bibliography.] Ernest Nagel, *The Structure of Science* (New York: Harcourt, 1961).

50. [According to Popper, preference should be given to theories "which can be most severely tested," which is "equivalent to a rule favoring theories with the highest possible empirical content." Karl Popper, *The Logic of Scientific Discovery* (New York: Routledge, 2002), 105.]

51. [The French manuscript gives "leur caractère gnoséologique particulier," which is here translated as "their epistemological nature."]

52. Edgar Morin, *Le paradigme perdu: la nature humaine* (Paris: Le Seuil, 1973).

53. Popper, *Logic of Scientific Discovery*, 8.

54. J. Hadamard, *An Essay on the Psychology of Invention in the Mathematical Field*, 2nd ed. (New York: Dover, 1954).

55. Édouard Claparède, *La genèse de l'hypothèse* (Geneva: Kundig, 1934).

56. K. Duncker, "A Qualitative (Experimental and Theoretical) Study of Productive Thinking (Solving of Comprehensible Problems)," *Journal of Genetic Psychology* 33 (1926).

57. Max Wertheimer, *Productive Thinking* (New York: Harper, 1945).

58. Jerome S. Bruner and Leo Postman, "On the Perception of Incongruity: A Paradigm," *Journal of Personality* 18 (1949).

59. Gruber, *Darwin on Man*.

60. [The published English version gives "political" whereas the French manuscript gives "ideological" ("*idéologique*," "*idéologie*").]

61. [Due to the difficulty in translating the French term "gnoséologique" in this context, I give the last sentence of the text in French: "Deuxièmement, la position gnoséologique de ce problème semble être tellement centrale qu'il représente le noeud gordien de toutes les approches possibles et qu'aucune méthode d'investigation unidisciplinaire ne donne, à elle seule, des résultats entièrement satisfaisants." Grmek, "Plaidoyer pour la démythification," 89.]

8. A MEMORICIDE

1. [Grmek is likely referring to the Museum of Croatian Archaeological Monuments that houses the largest collections of cultural items from the seventh to the fifteenth centuries and, in particular, from the early medieval Croatian state (i.e., from the ninth to the twelfth centuries). Also, according to the Museums Documentation Center (http://ratne-stete.mdc.hr/en/war-damages/), which collects information about war damages to museums in Croatia, this building is the only one listed for the city of Split.]

9. DUBROVNIK: THE SLAVIC ATHENS

1. [The separation between "producers" and "users" of knowledge could be a reference to Ludwik Fleck (1896–1961), a Polish Jewish bacteriologist and influential epistemologist who distinguished between "esoteric" and "exoteric" knowledge. See Ludwik Fleck, *Genesis and Development of a Scientific Fact*, trans.

Fred Bradley and Thaddeus J. Trenn (1935; Chicago: University of Chicago Press, 1979), 104–5.]

2. Ivan Gundulić, *Dubravaka* (1628; Dubrovnik, 1837).

3. [On Grmek's work on Boscovich, see especially Mirko D. Grmek, "La méthodologie de Boscovich," *Revue d'histoire des sciences* 49 (1996).]

4. Roger Joseph Boscovich, *A Theory of Natural Philosophy*, Latin-English ed. (1758; Chicago: Open Court, 1922).

5. Roger Joseph Boscovich, *Les éclipses. Poëme en six chants, dédié à sa Majesté*, trans. M. L'Abbé de Barruel (Paris: Valade and Laporte, 1779), 408–9 [my translation].

6. [Grmek's first article, published in 1946, was on Baglivi; see Introduction.]

7. Charles Daremberg, *Histoire des sciences médicales: Depuis Harvey jusqu'au XIXᵉ siècle*, vol. 2 (Paris: Baillière, 1870), 783.

8. Giorgio Baglivi, "Dissertatio VI. De Anatome, Morfu, & Effectibus Tarantulae," in *Opera Omnia. Medico-Pratica et Anatomica* (Lyon, France: Anisson, 1714), 639.

9. [For a comprehensive analysis of the development of quarantine measures against plague and other infectious diseases in Dubrovnik during the Middle Ages, see Zlata Blažina Tomić and Vesna Blažina, *Expelling the Plague: The Health Office and the Implementation of Quarantine in Dubrovnik, 1377–1533* (Montreal, Canada: McGill-Queen's University Press, 2015).]

10. [See Chapter 8 in this volume.]

Abderhalden, Emil. "Wandlungen in der Auffassung des Wesens der Alternsvor-
gänge." *Bulletin der Schweizerischen Akademie der Medizinischen Wissenschaften*
6 (1950): 102–14.
Ackerknecht, Erwin H. "Causes and Pseudocauses in the History of Diseases."
In *A Celebration of Medical History: The Fiftieth Anniversary of the Johns Hopkins
Institute of the History of Medicine and the Welch Medical Library*, edited by
Lloyd G. Stevenson, 19–31. Baltimore, Md.: Johns Hopkins University
Press, 1982.
———. *Geschichte und Geographie der wichtigsten Krankheiten*. Stuttgart, Ger-
many: Inke, 1963.
Alizon, Marc, and Luc Montagnier. "Genetic Variability in Human Immuno-
deficiency Viruses." *Annals of the New York Academy of Sciences* 511 (1987):
376–84.
Alizon, Samuel, A. Hurford, N. Mideo, and M. Van Baalen. "Virulence Evolu-
tion and the Trade-Off Hypothesis: History, Current State of Affairs and the
Future." *Journal of Evolutionary Biology* 22 (2009): 245–59.
Allison, Anthony C. "Protection Afforded by Sickle-Cell Trait against Subter-
tian Malarial Infection." *British Medical Journal* 1 (1954): 290–94.
Allyn, H. B. "The Black Death: Its Social and Economic Results." *Annals of
Medical History* 7 (1925): 226–36.
Ameisen, Jean-Claude. "Programmed Cell Death and AIDS Pathogenesis." In
Huitième Colloque des Cent-Gardes, edited by M. Girard and L. Valette, 5–10.
Lyon, France: Mérieux, 1993.
Anderson, Roy M. "The Ecological Factors That Determine the Impact of a
New Disease in Host Population." Paper presented at the International Sym-
posium on Emerging Infectious Diseases: Historical Perspectives, Annecy,
France, April 6–8, 1992.
Anselmi, Aurelio. *Gerocomica, sive, De senum regimine: opus non modo philosophis &
medis gratum, sed hominibus utile*. Venice: Apud Franciscum Ciottum, 1606.
Antiseri, Dario. "Prova di une teoria ed educazione al riconosciemento dell
'errore.'" *Medicina nei Secoli* 13 (1976): 307–64.

Aristotle. *On Youth and Old Age, Life and Death, and Respiration.* Translated by
W. Ogle. New York: Longmans, Green, 1902.

―――. *Politics.* Translated by H. Rackman. Cambridge, Mass.: Harvard University Press, 1932.

Arrizabalaga, Jon. "History of Disease and the Longue Durée." *History and Philosophy of the Life Sciences* 27 (2005): 41–56.

―――. "Problematizing Retrospective Diagnosis in the History of Disease." *Asclepio* 54 (2002): 51–70.

Ashburn, Percy M. *The Ranks of Death: A Medical History of the Conquest of America.* New York: Coward-McCann, 1947.

Aziz, Douglas C., Zaher Hanna, and Paul Jolicoeur. "Severe Immunodeficiency Disease Induced by a Defective Murine Leukemia Virus." *Nature* 338 (1989): 505–8.

Bachelard, Gaston. *The Formation of the Scientific Mind: A Contribution to a Psychoanalysis of Objective Knowledge.* Translated by Mary McAllester Jones. Manchester, UK: Clinamen, 2002.

Bacon, Francis. *Historia vitae et mortis.* London: Lownes, 1623.

―――. *Sylva Sylvarum, or A Natural History in Ten Centuries. Whereunto Is Newly Added the History Natural and Experimental of Life and Death, or of the Prolongation of Life.* London: J. R., 1650.

Bacopoulos-Viau, Alexandra, and Aude Fauvel, eds. "Tales from the Asylum: Patient Narratives and the (De)Construction of Psychiatry." Special issue, *Medical History* 60, no. 1 (2016).

Bacq, Zenon M. *Les transmissions chimiques de l'influx nerveux.* Paris: Gauthier-Villars, 1974.

Baglivi, Giorgio. "Dissertatio VI. De Anatome, Morfu, & Effectibus Tarantulae." In *Opera Omnia. Medico-Pratica et Anatomica.* Lyon, France: Anisson, 1714.

―――. "De praxi medica." In *Opera omnia medico-pratica et anatomica.* Venice, 1727.

Baker, B. J., and G. J. Armelagos. "The Origin and Antiquity of Syphilis." *Current Anthropology* 29 (1988): 703–37.

Baldini, Massimo. *Epistemologia e storia della scienza.* Florence: Città di Vita, 1974.

Bardinet, Thierry. "Le cœur et le système des 'conduits' (metou) dans les conceptions physiologiques de l'Égypte ancienne." PhD diss., Université de Paris, EPHE, 1992.

―――. "Remarques sur les maladies de la peau, la lèpre et le châtiment divin dans l'Égypte ancienne." *Revue d'Égyptologie* 39 (1988): 3–36.

Barthez, Paul-Joseph. *Nouvelle méchanique des mouvements de l'homme et des animaux.* Paris: Méquignon l'aîné, an VI, 1798.

Benjamin, H. "Biologic versus Chronologic Age." *Journal of Gerontology* 2 (1947): 217–27.

Bergh, Øivind, Knut Yngve Børsheim, Gunnar Bratbak, and Mikal Heldal. "High Abundance of Viruses Found in Aquatic Environments." *Nature* 340 (1989): 467.

Bernard, Claude. *Cahier de notes 1850–1860. Présenté et commenté par Mirko D. Grmek.* with a preface by Robert Courrier. Paris: Gallimard, 1965.

———. *Claude Bernard and Experimental Medicine: Collected Papers from a Symposium Commemorating the Centenary of the Publication of* An Introduction to the Study of Experimental Medicine *and the First English Translation of Claude Bernard's* Cahier Rouge, edited by Francisco Grande and Maurice B. Visscher. Cambridge, Mass.: Schenkman, 1967.

———. "De l'origine du sucre dans l'économie animale." *Archives générales de médecine* 4 (1848): 303–19.

———. "Du suc gastrique et de son rôle dans la nutrition." Thesis for the doctorate in medicine. Paris: Rignoux, 1843.

———. *Introduction à l'étude de la médecine expérimentale.* Paris: Baillière, 1865.

———. *An Introduction to the Study of Experimental Medicine.* Translated by H. Copley Greene. New York: Schuman, 1949.

———. *Leçons sur le diabète.* Paris: Baillière, 1877.

———. "The Origin of Sugar in the Animal Body." In *Medical Classics.* Vol. 3, edited by Emerson Crosby Kelly, 567–80. Baltimore, Md.: Williams and Wilkens, 1939.

———. *Philosophie. Manuscrit inédit. Texte présenté par Jacques Chevalier.* Paris: Hatier-Boivin, 1937.

———. "Recherches sur une nouvelle fonction du foie considéré comme producteur de matière sucrée chez l'homme et les animaux." Thesis for the degree of doctor of natural science. Paris: Martinet, 1853.

Bernard, Claude, and Charles Barreswil. "De la présence du sucre dans le foie." *Compte Rendus de l'Académie des Sciences* 27 (1848): 514–15.

———. "Du sucre dans l'œuf." *Compte Rendus de la Société de Biologie* 1 (1849): 64.

Bernard, Jean. "Esquisse d'une géographie des maladies du sang." *Annales de géographie* 74 (1965): 271–80.

Bernard, Jean, and Jacques Ruffié. *Hématologie géographique.* Paris: Masson, 1966.

Bett, Walter R. *The History and Conquest of Common Diseases.* Norman: University of Oklahoma Press, 1954.

Beveridge, William I. *Influenza: The Last Plague. An Unfinished Story of Discovery.* New York: Prodist, 1977.

Bhushan, B. "AIDS: A Soviet Propaganda Tool." *The Times of India,* November 19, 1986.

Biagi, Enzo. *Il sole malato. Viaggio nella paura dell'AIDS.* Milan: Mondadori, 1987.

Bickel, Marcel H., ed. *Correspondence Henry E. Sigerist to Richard H. Shryock (1933–1956).* Self-published, 2012.

Binet, L., and F. Bourlière. *Précis de gérontologie.* Paris: Masson, 1955.

Biraben, Jean-Noël. *Les hommes et la peste dans les pays européens et méditerranéens.* Paris: Moudon, 1975–1976.

Blanc, Georges. "La disparition de la peste et ses causes épidémiologiques." *Semaine des hôpitaux* 37 (1961): 105–10.

Blažina Tomić, Zlata, and Vesna Blažina. *Expelling the Plague: The Health Office and the Implementation of Quarantine in Dubrovnik, 1377–1533*. Montreal, Canada: McGill-Queen's University Press, 2015.

Bogomolets, Alexander A. *Prodlenie zhini*. Kiev, 1938.

———. *The Prolongation of Life*. Translated by P. V. Karpovich and S. Bleeker. New York: Sloan and Pearce, 1946.

Borelli, Giovanni Alfonso. *De motu animalium*. Rome: Angeli Bernabò, 1680–1681.

Boscovich, Roger Joseph. *Les éclipses. Poëme en six chants, dédié à sa Majesté*. Translated by M. L'Abbé de Barruel. Paris: Valade and Laporte, 1779.

———. *A Theory of Natural Philosophy*. Latin-English Edition. Chicago: Open Court, 1922.

Bowsky, William M. "The Impact of the Black Death upon Sienese Government and Society." *Speculum* 39 (1964): 1–34.

Boyden, Stephen. *Western Civilization in Biological Perspective: Patterns in Biohistory*. Oxford: Clarendon, 1987.

Brachet, Auguste. *Pathologie mentale des rois de France. Louis XI et ses ascendants*. Paris: Hachette, 1903.

Brandt, Allan M. "Emerging Themes in the History of Medicine." *Milbank Quarterly* 69 (1991): 199–214.

Brau, Paul. *Trois siècles de médecine coloniale française*. Paris: Vigot, 1931.

Braunstein, Jean-François. "Entretien avec Jean-François Braunstein." *Intelligere, Revista de História Intelectual* 2 (2016): 172–85.

Bremermann, H. J., and Roy M. Anderson. "Mathematical Models of HIV Infection: Threshold Condition for Transmission and Host Survival." *Journal of Acquired Immune Deficiency Syndromes* 3 (1990): 1129–34.

Brès, P. "Les virus Lassa, Marburg et Ebola, nouveaux venus en pathologie tropicale." *Nouvelle Presse Médicale* 7 (1978): 2921–26.

Brieger, Gert H. "The Historiography of Medicine." In *Companion Encyclopedia of the History of Medicine*, edited by William F. Bynum and Roy Porter, 22–44. London: Routledge, 1993.

Brossolet, J. "Expansion européenne de la suette anglaise." In *Proceedings of the XIII International Congress of the History of Medicine, 1972* (1974): 595–600.

Brousseau, Jérôme. "La sénescence et la mort des individus sont-elles utiles à l'espèce? Regards sur la théorie évolutionniste du vieillissement de Mirko D. Grmek." In *Médecine, science, histoire. Le legs de Mirko Grmek*, edited by Pierre-Olivier Méthot. Paris: Matériologiques, forthcoming.

Brown-Séquard, Charles E. "Des effets produits chez l'homme par des injections sous-cutanées d'un liquide retiré des testicules frais de cobayes et de chiens." *Compte rendus de la Société de Biologie* 41 (1889): 415–22.

Bruner, Jerome S., and Leo Postman. "On the Perception of Incongruity: A Paradigm." *Journal of Personality* 18 (1949): 206–23.

Buklijas, Tatjana. "Dalmacija u djelu Mirka Dražena Grmeka (1924–2000)." *Građa I prilozi za povijest Dalmacije* 16 (2000): 381–91.

———. "Dubrovnik in the Work of Mirko Dražen Grmek." *Dubrovkik Annals* 5 (2001): 119–23.

Bürger, M. *Altern und Krankheit*. 3rd ed. Leipzig, Germany: Georg-Thieme, 1957.

Burnet, Frank Macfarlane. *Viruses and Man*. Harmondsworth, England: Penguin, 1955.

Bütschli, Otto. "Gedanken über Leben und Tod." *Zoologischer Anzeiger* 5 (1882): 64–67.

Cabanès, Augustin. *L'histoire éclairée par la clinique*. Paris: Albin Michel, 1921.

Canguilhem, Georges. *Knowledge of Life*. Translated by Stefanos Geroulanos and Daniela Ginsburg. New York: Fordham University Press, 2008.

———. "Machine and Organism." In *Knowledge of Life*, 75–97. New York: Fordham University Press, 2008.

———. *The Normal and the Pathological*. Translated by Carolyn R. Fawcett in collaboration with Robert S. Cohen, with an introduction by Michel Foucault. New York: Zone Books, 1991.

———. "The Object of the History of Sciences." In *Continental Philosophy of Science*, edited by Garry Gutting, 198–207. Malden, Mass.: Blackwell, 2005.

Cannon, Walter, B. *The Way of an Investigator: A Scientist's Experiences in Medical Research*. New York: Norton, 1945.

Carmichael, Ann G., and Arthur M. Silverstein. "Smallpox in Europe before the Seventeenth Century: Virulent Killer or Benign Disease?" *Journal of the History of Medicine and Allied Sciences* 42 (1987): 147–68.

Carpentier, Elisabeth. "Autour de la peste noire: famines et épidémies dans l'histoire du XIVᵉ siècle." *Annales: Économies, Sociétés, Civilisations* 17 (1962): 1062–92.

———. *Une ville devant la peste, Orvieto et la peste noire de 1348*. Paris: S.E.V.P.E.N., 1963.

Carrel, Alexis. *Man: The Unknown*. New York: Harper and Brothers, 1935.

———. "Physiological Time." *Science* 74 (1931): 618–21.

———. "Rejuvenation of Culture Tissues." *Journal of American Association of Medicine* 57 (1911): 1611.

Carrel Alexis, and A. H. Ebeling. "Antagonistic Growth Principles of Serum and Their Relation to Old Age." *Journal of Experimental Medicine* 38 (1923): 419–25.

Carrel, Alexis, and Charles A. Lindbergh. *The Culture of Organs*. New York: Hoeber, 1938.

Carter, Henry R. *Yellow Fever: An Epidemiological and Historical Study of Its Place of Origin*. Baltimore, Md.: Williams and Wilkins, 1931.

Carter, Simon, John L. Allen, Nick Bingham, and Steve Hinchliffe. *Pathological Lives: Disease, Space, and Biopolitics*. Chichester, England: Wiley, 2016.

Chaussinand, Roland. "Tuberculose et lèpre, maladies antagoniques. Éviction de la lèpre par la tuberculose." *International Journal of Leprosy* 16 (1948): 431–38.

Claparède, Édouard. *La genèse de l'hypothèse*. Geneva: Kundig, 1934.

Clark, S. P., and T. W. Mak. "Fluidity of a Retrovirus Genome." *Journal of Virology* 50 (1984): 759–65.

Cockburn, Aidan. *The Evolution and Eradication of Infectious Diseases*. Baltimore, Md.: Johns Hopkins University Press, 1963.

Colnat, Albert. *Les épidémies et l'histoire*. Paris: Le François, 1937.

Comfort, Alexis. *The Biology of Senescence*. New York: Rinehart, 1956.

Corbellini, Gilberto, and Chiara Preti. "Towards an Evolutionary Historiography and Epistemology of Medicine: The Legacy of Mirko Grmek." *Medicina Nei Secoli* 20 (2008): 965–83.

Corradi, Alfonso. *Annali delle epidemie occorse in Italia delle prime memorie fino al 1850*. 4 vols. Bologna: Gamberini and Parmeggiani, 1865–1880.

Coste, Joël, Bernardino Fantini, and Louise L. Lambrichs, eds. *Le concept de pathocénose de Mirko Grmek. Une conceptualisation novatrice de l'histoire des maladies*. Geneva: Droz, 2016.

Coury, Charles. *Grandeur et déclin d'une maladie. La tuberculose au cours des âges*. Suresnses, France: Lepetit, 1972.

Coville, A. "Écrits contemporains sur la peste de 1348 à 1350." *Histoire littéraire de la France* 37 (1928): 325–90.

Cowley, G. "The Future of AIDS." *Newsweek* 22 (1993): 45–50.

Cox, F. E. G. "The Worm and the Virus." *Nature* 347 (1990): 618.

Creighton, Charles. *A History of Epidemics in Britain*. London: Cambridge University Press, 1891.

Crick, Francis. *Life Itself: Its Origin and Nature*. New York: Simon and Schuster, 1981.

Crosby, Alfred W. *The Columbian Exchange: Biological and Cultural Consequences of 1492*. Westport, Conn.: Greenwood, 1972.

———. "Conquistador y Pestilencia: The First New World Pandemic and the Fall of the Great Indian Empires." *The Hispanic American Historical Review* 47 (1967): 321–37.

Curtin, Philip D. *Death by Migration: Europe's Encounter with the Tropical World in the Nineteenth Century*. New York: Cambridge University Press, 1989.

Dan, Bruce B. "Toxic Shock Syndrome: Back to the Future." *The Journal of the American Medical Association* 257 (1987): 1094–95.

Daremberg, Charles. *Histoire des sciences médicales: Depuis Harvey jusqu'au XIXe siècle*. Vol. 2. Paris: Baillière, 1870.

Darwin, Charles. *On the Origins of Species by Means of Natural Selection, or the Preservation of Favoured Races in the Struggle for Life*. London: Murray, 1859.

Dawkins, Richard. *The Selfish Gene*. Oxford: Oxford University Press, 1976.

de Buzon, Catherine [now Chevalley]. "Remarques sur l'interprétation de l'œuvre de Kepler." *Archives Internationales d'histoire des sciences* 27 (1977): 72–81.

De Leys, Robert, Bart Vanderborght, Marleen Vanden Haesevelde, Leo Hendrickx, Anja van Geel, Carlo Wauters, Ria Bernaerts et al. "Isolation and Partial Characterization of an Unusual Immunodeficiency Retrovirus from

Two Persons of West-Central-Africa Origin." *Journal of Virology* 64 (1990): 1207–16.

de Romo, Ana Cecilia Rodríguez. "Eloge: Mirko D. Grmek, 9 January 1924–6 March 2000." *Isis* 92 (2001): 742–44.

Descartes, René. *Discourse on the Method of Rightly Conducting the Reason and Seeking the Truth in the Sciences.* In *The Philosophical Writings of Descartes.* Vol. 1. Translated by John Cottingham, Robert Stoothoff, and Dugald Murdoch. Cambridge: Cambridge University Press, 1985.

———. "Letter to Henry More, February 5, 1649." In *The Philosophical Writings of Descartes: The Correspondence.* Vol. 3. Translated by John Cottingham, Robert Stoothoff, Dugald Murdoch, and Anthony Kenny, 360–67. Cambridge: Cambridge University Press, 1991.

———. *Treatise on Man.* In *The Philosophical Writings of Descartes.* Translated by John Cottingham, Robert Stoothoff, and Dugald Murdoch. Cambridge: Cambridge University Press, 1985.

Devignat, René. *La peste antique du Congo belge dans le cadre de l'histoire et de la géographie.* Brussels: Institut Royal Colonial Belge, 1953.

———. "Répartition géographique des trois variétés de *Pasteurella pestis.*" *Schweizerische Zeitschrift für Allgemeine Pathologie* 16 (1953): 509–15.

———. "Variétés de l'espèce *Pasteurella pestis*; nouvelle hypothèse." *Bulletin of the World Health Organization* 4 (1951): 247–63.

Dhar, N. R. "Old Age and Death from a Chemical Point of View." *Journal of Physical Chemistry* 30 (1926): 378–82.

Dietrich, Ursula, Michalina Adamski, Renate Kreuz, Anja Seipp, Herbert Kühnel, and Helga Rübsamen-Waigmann. "A Highly Divergent HIV–2 Related Isolate." *Nature* 342 (1989): 948–50.

Dijksterhuis, Eduard J. "The Origins of Classical Mechanics from Aristotle to Newton." In *Critical Problems in the History of Science*, edited by Marshall Clagett, 163–84. Madison: Wisconsin University Press, 1969.

Divry, P. "Considération sur le vieillissement cérébral." *Journal belge de neurologie et de psychiatrie* 47 (1947): 65–81.

Dobzhansky, Theodosius. *Mankind Evolving: The Evolution of the Human Species.* New Haven, Conn.: Yale University Press, 1962.

Dodin, A. "Pourquoi les pandémies cholériques?" *Bulletin de la Société de pathologie exotique* 77 (1984): 127–34.

Downs, Wilbur G. "History of Epidemiological Aspects of Yellow Fever." *Yale Journal of Biology and Medicine* 55 (1982): 179–85.

Doyon, André, and Lucien Liaigre. *Jacques Vaucasson, mécanicien de génie.* Paris: Presses Universitaires de France, 1966.

———. "Méthodologie comparée du biomécanisme et de la mécanique comparée." *Dialectica* 10 (1956): 292–323.

Dubos, René J. *The White Plague: Tuberculosis, Man and Society.* Boston: Little, 1952.

Dubos, René J., and James. G. Hirsch. *Bacterial and Mycotic Infections of Man.* 4th ed. Philadelphia, Penn.: Lippincott, 1965.

Duesberg, P. H. "Human Immunodeficiency Virus and Acquired Immunodeficiency Syndrome: Correlation but Not Causation." *Proceedings of the National Academy of Sciences* 86 (1989): 755–64.

―――. "Retroviruses as Carcinogens and Pathogens: Expectations and Reality." *Cancer Research* 47 (1987): 1199–1220.

Duffin, Jacalyn. "Best-Intentioned Efforts Aggravated AIDS Pandemic." *Canadian Medical Association Journal* 184 (2012): E981–E982.

―――. "In Memoriam: Mirko Dražen Grmek." *Bulletin of the History of Medicine* 74 (2000): 561–65.

―――. *Lovers and Livers: Disease Concepts in History*. Toronto: Toronto University Press, 2002.

Dugac, Želijko. "New Public Health for a New State: Interwar Public Health in the Kingdom of Serbs, Croats, and Slovenes and the Rockefeller Foundation." In *Facing Illness in Troubled Times: Health in Europe in the Interwar Years, 1918–1939*, edited by Iris Borowy and Wolf D. Gruner, 277–304. Frankfurt, Germany: Lang, 2005.

Duić, Vjera. "Overview." In *The History of East-Central European Eugenics, 1900–1945*, edited by Marius Turda, 422–35. London: Bloomsbury, 2015.

Dumas, Jean-Baptiste, and Jean Baptiste Boussingault. *The Chemical and Physiological Balance of Organic Nature: An Essay*. New York: Saxton, 1844.

―――. *Essai de statique chimique des êtres organisés*. 3rd ed. Paris: Masson, 1844.

Duncker, K. "A Qualitative (Experimental and Theoretical) Study of Productive Thinking (Solving of Comprehensible Problems)." *Journal of Genetic Psychology* 33 (1926): 642–708.

Eigen, M., and P. Schuster. *The Hypercycle: A Principle of Natural Self-Organization*. New York: Springer, 1979.

Eigern, N. *Die Stufen zum Leben*. Munich, Germany: Pipper, 1987.

Espinas, Alfred. "L'organisme ou la machine vivante en Grèce au IV\ siècle avant J. C." *Revue de métaphysique et de morale* 11 (1903): 703–15.

Esser, Albert. *Cäsar und die julisch-claudischen Kaiser in biologisch-ärztlichen Blickfeld*. Leiden, Netherlands: Brill, 1958.

Essex, M. "Origin of AIDS." In *AIDS: Etiology, Diagnosis, Treatment and Prevention*, edited by Vincent T. DeVita Jr., Samuel Hellman, and Steven A. Rosenberg, 3–11. 3rd ed. Philadelphia, Penn.: Lippincott, 1993.

Essex, M., and P. J. Kanki. "The Origin of the AIDS Virus." *Scientific American* 259 (1988): 64–71.

Ewald, Paul. "Transmission Modes and the Evolution of Virulence with Special Reference to Cholera, Influenza and AIDS." *Human Nature* 2 (1990): 1–30.

Fantini, Bernardino. "La biologica molecolare e il concetto di malattia." In *Maladie et maladies, histoire et conceptualisation (Mélanges en l'honneur de Mirko Grmek)*, edited by Danielle Gourevitch, 449–86. Geneva: Droz, 1992.

―――. "Obituary: Mirko Dražen Grmek (1924–2000)." *Medical History* 45 (2001): 273–79.

———. "Social and Biological Origins of the AIDS Pandemic." In *AIDS and the Historian*, edited by Victoria A. Harden and Guenter B. Risse, 52–55. Bethesda, Md.: NIH, 1991.

Fatović-Ferenčić, Stella. "'Society as Organism': Metaphors as Departure Point of Andrija Štampar's Health Ideology." *Croatian Journal of Medical History* 49 (2008): 709–19.

Fauci, Anthony S. "Multifactorial and Multiphasic Components of the Immunopathogenic Mechanisms of HIV Disease." In *Huitième Colloque des Cent-Gardes*, edited by M. Girard and L. Valette, 81–85. Lyon, France: Mérieux, 1993.

Fenner, Frank et al. *The Biology of Animal Viruses*. New York: Academic Press, 1974.

Feyerabend, Paul. *Against Method: Outline of an Anarchist Theory of Knowledge*. London: New Left Books, 1975.

Fields, B. N. *Virology*. New York: Raven, 1990.

Finke, Leonhard L. *Versuch einer allgemeinen medicinisch-praktischen Geographie*. 3 vols. Leipzig, Germany: Weidmann, 1792–1795.

Fisher, Ronald A. "Has Mendel's Work Been Rediscovered?" *Annals of Science* 1 (1936): 115–37.

Fleck, Ludwik. *Genesis and Development of a Scientific Fact*. Translated by Fred Bradley and Thaddeus J. Trenn. Chicago: University of Chicago Press, 1979.

Floch, H. "La réaction de Mitsuda rendue positive par une primo-infection tuberculeuse est-elle accompagnée d'une immunité relative antilépreuse?" *Bulletin de la Société de pathologie exotique* 47 (1954): 771–75.

Florkin, Marcel. *Naissance et déviation de la théorie cellulaire dans l'œuvre de Th. Schwann*. Paris: Hermann, 1960.

Folke, Henschen. *The History and Geography of Diseases*. New York: Delacorte, 1966.

———. "La nature réelle du vieillissement." *Revue médicale de Bruxelles* 33 (1953): 2061–68.

Fontenay, F., and L. De Villepin. "Un entretien avec le Pr. Luc Montagnier." *Journal du Sida* 52–53 (1993): 20–21.

Fornaciari, Gino, Mauro Castagna, Adele Togneti, Daniela Tornaboni, and Joseph Bruno. "Syphilis in a Renaissance Italian Mummy." *The Lancet* 334 (1989): 614.

Foucault, Michel. "Cuvier's Situation in the History of Biology." *Foucault Studies* 22 (2017): 208–37.

Frame, John D., John M. Baldwin Jr., David J. Gocke, and Jeanette M. Troup. "Lassa Fever: A New Virus Disease of Man from West Africa." *The American Journal of Tropical Medicine and Hygiene* 19 (1970): 670–76.

Fraser, David W., Theodore R. Tsai, Walter Orenstein, William E. Parkin, James Beecham, Robert G. Sharrar, John Harris et al. "Legionnaire's Disease: Description of an Epidemic of Pneumonia." *New England Journal of Medicine* 297 (1977): 1189–97.

Fribourg-Blanc, André, and Henri H. Mollaret. "Natural Treponematosises of the African Primate." *Primates in Medicine* 3 (1968): 110–18.

Fribourg-Blanc, André, G. Niel, and Henri H. Mollaret. "Confirmation sérologique et microscopique de la tréponématose du cynocéphale de Guinée." *Bulletin de la Société de pathologie exotique* 59 (1966): 54–59.

Fultz, Patricia N., Harold M. McClure, Daniel C. Anderson, and William M. Switzer. "Identification and Biologic Characterization of an Acutely Lethal Variant of Simian Immunodeficiency Virus from Sooty Mangabeys (SIV/SMM) AIDS." *AIDS Research and Human Retroviruses* 5 (1989): 397–409.

Galen. *Adversus Lycum Libellus*. In *Galeni adversus Lycum et adversus Iulianum libelli*. Corpus mediocorum Graecorum, vol. 7, edited by E. Wenkebach. Berlin: Verlag, 1951.

———. *Claudii Galeni, Opera omnia*, edited by Carl G. Kühn. Leipzig, Germany: Cnobloch, 1921.

———. *On the Usefulness of the Parts of the Body*. Translated by Margaret Tallmadge May. Ithaca, N.Y.: Cornell University Press, 1968.

———. *Selected Works*. Translated by Peter N. Singer. Oxford: Oxford University Press, 1997.

Gallo, Robert. *Virus Hunting*. New York: Basic Books, 1991.

Gasquet, Cardinal F. A. *The Black Death of 1348 and 1349*. London: Bell, 1893.

Gayon, Jean. "L'Institut d'histoire des sciences." *Cahiers Gaston Bachelard* 14 (2016): 15–63.

Geison, Gerald L. *The Private Science of Louis Pasteur*. Princeton, N.J.: Princeton University Press, 1995.

Gelfand, Toby. "The Annales and Medical Historiography: *Bilan et perspectives*." In *Problems and Methods in the History of Medicine*, edited by Roy Porter and Andrew Wear, 15–39. London: Croom-Helm, 1987.

Geymonat, Ludovico. *Galileo Galilei*. 2nd ed. Turin, Italy: Einaudi, 1962.

Gilbert, Judson B., and Gordon E. Mestler. *Disease and Destiny: A Bibliography of Medical References to the Famous*. London: Dawsons, 1962.

Gilbert, N. P. *Histoire médicale de l'armée française à Saint-Domingue en 1802*. Paris: Gabon et Cie, 1803.

Gilson, Étienne. "Descartes, Harvey et la scolastique." In *Études de philosophie médiévale*, 191–245. Strasbourg, France: Palais de l'Université, 1921.

Girard, Rolande. *Tristes chimères. Sida*. Paris: Grasset, 1987.

Goldzreher, M. A. "Endocrine Aspects of Senescence." *Geriatrics* 1(1946): 226–31.

Goodenough, M., T. Huet, W. Saurin, S. Kwok, J. Sninsky, and S. Wain-Hobson. "HIV–1 Isolates Are Rapidly Evolving Quasispecies: Evidence for Viral Mixture and Preferred Nucleotide Substitutions." *Journal of Acquired Immune Deficiency Syndromes* 2 (1989): 344–52.

Gould, Peter R. *The Slow Plague: A Geography of the AIDS Pandemic*. Cambridge, Mass.: Blackwell, 1993.

Goupil, J. M. "L'histoire de la coqueluche." PhD diss., University of Caen, 1976.

Gourevitch, Danielle. "Biographie et bibliographie choisie." In *Maladie et mala-dies. Histoire et conceptualization (Mélanges en l'honneur de Mirko Grmek)*, edited by Danielle Gourevitch, xlvi–lviii. Paris: Champion, 1991.

———. "Il est en Grèce une divinité . . . ." *Histoire des sciences médicales* 35 (2001): 347–49.

Gourevitch, Danielle, and Mirko D. Grmek. *Les maladies dans l'art antique*. Paris: Fayard, 1998.

Granger, J. P. "The Plague as a Factor in History." *Glasgow Medical Journal* 77 (1912): 178–86, 260–73.

Great, Frederick the. "Eulogy on Julien Offroy de La Mettrie." In Julien Offroy de La Mettrie, *Man a Machine*. Translated by Gertrude C. Bussey, revised by M. W. Calkins, 1–9. Chicago: Open Court, 1912.

Green, Monica, H. "The Value of Historical Perspective." In *Research Companion to the Globalization of Health*, edited by Ted Schrecker, 17–37. Aldershot, England: Ashgate, 2012.

Greenough, A., and J. A. Davis. "*Encephalitis lethargica*: Mystery of the Past or Undiagnosed Disease of the Present?" *The Lancet* 1 (1983): 922–23.

Griffin, J. J. "Aristotle's Observations on Gerontology." *Geriatrics* 5 (1950): 222–26.

Grmek, Mirko D. "Ancient Slavic Medicine." *Journal of the History of Medicine and Allied Sciences* 14, no. 1 (1959): 18–40.

———. "Aux États-Unis, un colloque international sur les rapports entre la Bio-logie, l'Histoire et la Philosophie, Denver, 27 novembre–2 décembre 1967." *Revue d'histoire des sciences* 21 (1968): 169–71.

———. *Catalogue des manuscrits de Claude Bernard. Avec la bibliographie de ses travaux imprimés et des études sur son œuvre*. With a foreword by M. Bataillon and E. Wolff and with an introduction by L. Delhoume and P. Huard. Paris: Collège de France and Masson, 1967.

———. "The Concept of Disease." In *Western Medical Thought from Antiquity to the Middle Ages*, edited by Mirko D. Grmek, coordinated by Bernardino Fan-tini, 241–258. Translated by Antony Shugaar. Cambridge, Mass.: Harvard University Press, 1998.

———. "Dark Sides of the Viral Causal Explanation of AIDS." *Croatian Medical Journal* 35 (1994): 1–18.

———. "Définition du domaine propre de l'histoire des sciences et considéra-tions sur ses rapports avec la philosophie des sciences." *History and Philosophy of the Life Sciences* 1 (1979): 3–12.

———. "Definitions." In Mirko D. Grmek, *On Ageing and Old Age: Basic Prob-lems and Historic Aspects of Gerontology and Geriatrics*, 3–7. The Hague: W. Junk, 1958.

———. "Discours de Rome." Paper presented at Personnalités, problèmes et méthodes de l'histoire des sciences. Cinquante ans d'une discipline entre science de l'homme et science de la nature, Rome, Istituto della Enciclopedia Italiana, June 4–6, 1986.

———. "Discussion on Medicine and Culture." In *Medicine and Culture*, edited by Frederick N. L. Poynter, 48–50, 52–53, 119–20. London: Wellcome Institute, 1969.

———. *Diseases in the Ancient Greek World*. Translated by Mireille Muellner and Leonard Muellner. Baltimore, Md.: Johns Hopkins University Press, 1989.

———. "Entretien avec Mirko Grmek. Sida: Histoire d'une épidémie." *L'Histoire* 150 (1991): 50–55.

———. "Épidémiologie de la peste et histoire démographique et sociale." In *L'histoire des sciences et des techniques doit-elle intéresser les historiens? Colloque organisé par la Société française d'histoire des sciences et des techniques, 8 et 9 mai 1981*, 169–86. Paris: Société française d'histoire des sciences et des techniques, 1982.

———. "Évolution des conceptions de Claude Bernard sur le milieu intérieur." In *Philosophie et méthodologie scientifique de Claude Bernard*, 117–50. Paris: Masson, 1967.

———. "Examen critique de la genèse d'une grande découverte: la piqûre diabétique de Claude Bernard." *Clio Medica* 1 (1966): 341–50.

———. "Géographie médicale et histoire des civilisations." *Annales: Économies, Sociétés, Civilisations* 28 (1963): 1071–97.

———. "Histoire des recherches sur les relations entre le génie et la maladie." *Revue d'histoire des sciences* 15 (1962): 1071–97.

———. "Histoire des sciences et psychogenèse." *Cahiers de la Fondation Archives Jean Piaget* 4 (1983): 26–29, 87–91, 314–17, 409–13.

———. *Histoire du sida. Début et origine d'une pandémie actuelle*. Paris: Payot, 1989.

———. *History of AIDS: Emergence and Origin of a Modern Pandemic*. Translated by Russell C. Maulitz and Jacalyn Duffin. Princeton, N.J.: Princeton University Press, 1990.

———. Introduction to *Histoire de la pensée médicale en Occident*. Vol.1, *Antiquité et Moyen Age*, edited by Mirko D. Grmek, in collaboration with Bernardino Fantini, 7–24 (Paris: Seuil, 1995).

———. Introduction to *Western Medical Thought from Antiquity to the Middle Ages*, edited by Mirko D. Grmek, in collaboration with Bernardino Fantini, 1–21. Translated by Antony Shugaar. Cambridge, Mass.: Harvard University Press, 1998.

———. "La conception de la maladie et de la santé chez Claude Bernard." In *Mélanges Alexandre Koyré*. Vol. 2, *L'Aventure de la science*, 208–27. Paris: Hermann, 1964.

———. "La dénomination latine des maladies considérées comme nouvelles par les auteurs antiques." In *Le Latin médical (Mémoires du Centre Jean-Palerne, X)*, edited by Guy Sabbah, 195–214. Saint-Étienne, France: Presses de l'Université de Saint-Étienne, 1991.

———. "La gérontologie d'hier à aujourd'hui." In *Deux cents ans de gérontologie*, 7–10. Paris: La Société de gérontologie d'Ile de France, 1990.

———. *La guerre comme maladie sociale et autres textes politiques*. Paris: Seuil, 2001.

———. "La méthodologie de Boscovich." *Revue d'histoire des sciences* 49 (1996): 379–400.

———. "La notion de fibre vivante chez les médecins de l'école iatrophysiques." *Clio Medica* 5 (1970): 297–318.

———. "La paléopathologie des tumeurs osseuses malignes. Proposition d'une classification à l'usage de l'ostéo-archéologie, revue des exemples publiés et présentations de deux cas inédits." *Histoire des sciences médicales* 9 (1975): 21–50.

———. *La première révolution biologique. Réflexions sur la physiologie et la médecine au XVIIe siècle.* Paris: Payot, 1990.

———. "La réalité nosologique au temps d'Hippocrate." In *La collection hippocratique et son rôle dans l'histoire de la médecine (Strasbourg, 1972)*, edited by Mirko D. Grmek, 237–55. Leiden, Netherlands: Brill, 1975.

———. "La révolution biomédicale du XXe siècle." In *Histoire de la pensée médicale en Occident*, vol. III, edited by Mirko D. Grmek, 319–36. Paris: Seuil, 1999.

———. "La science chez les Slaves du Moyen-Age." In *La science antique et médiévale. Des origines à 1450*, edited by René Taton, 557–67. 2nd ed. Paris: Presses universitaires de France, 1966.

———. *La vie, les maladies et l'histoire.* Paris: Seuil, 2001.

———. "Le concept de maladie." In M. D. Grmek, *La vie, les maladies et l'histoire*, 21–28. Paris: Seuil, 2001.

———. "Le concept de maladie émergente." *History and Philosophy of the Life Sciences* 15 (1993): 281–96.

———. "L'émergence de la médecine scientifique en France sous le règne de Louis XIV." *Medizinhistorishes Journal* 11 (1976): 271–98.

———. "Le passé n'existe pas." L'actualité Poitou-Charentes, December (1997): 30–31.

———. "Le rôle du hasard dans la genèse des découvertes scientifiques." *Medicina nei Secoli* 13 (1976): 277–305.

———. "Les aspects historiques des problèmes fondamentaux de la gérontologie." *Le Scalpel* 110 (1957): 158–64.

———. "Le sida est-il une maladie nouvelle?" *Médecine et maladies infectieuses* 18 (1988): 577–82.

———. "Les origines d'une maladie d'autrefois: le scorbut des marins." *Bulletin de l'Institut océanographique de Monaco*, special issue 2 (1968): 505–23.

———. "L'étude du corps humain." In *La science moderne. De 1450 à 1800*, edited by René Taton, 140–54. 2nd ed. Paris: Presses Universitaires de France, 1969.

———. "Le vieillissement et la mort." In *Biologie*, edited by J. Rostand and A. Tétry, 779–827. Paris: Encyclopédie de la Pléiade, 1965.

———. "L'ex-Yougoslavie: la guerre comme maladie sociale." In *La guerre comme maladie sociale et autres textes politiques*, 99–115. Paris: Seuil, 2001.

———. "Médecine et épistémologie: transformation du savoir sur la santé et la maladie." *History and Philosophy of the Life Sciences* 10 (1987): 3–6.

———. "Medicinska škole u Dalmaciji u vrijeme francuske vladavine, 1806–1813" (PhD diss., University of Zagreb, 1958).

———. *On Ageing and Old Age: Basic Problems and Historic Aspects of Gerontology and Geriatrics.* The Hague: W. Junk, 1958.

———. "A Plea for Freeing the History of Scientific Discoveries from Myth." In *On Scientific Discovery: The Erice Lectures 1977*, edited by Mirko D. Grmek, Robert S. Cohen, and Guido Cimino, 9–42. Boston: Reidel, 1981.

———. "Préliminaires d'une étude historique des maladies." *Annales: Économie, Sociétés, Civilisations* (1969): 1437–83.

———. "Problèmes des maladies nouvelles." In *Sida, épidémies et sociétés*, 97–107. Lyon, France: Fondation Marcel Mérieux, 1987.

———. "Prolégomènes à une histoire générale des sciences." *Annales. Économies, Sociétés, Civilisations* 20 (1965): 138–46.

———. "Quelques notes intimes de Claude Bernard." *Archives Internationales d'histoire des sciences* 65 (1963): 339–52.

———. *Raisonnement expérimental et recherches toxicologiques chez Claude Bernard.* Geneva: Droz, 1973.

———. "Regard d'un historien sur les maladies émergentes." In *Proceedings of the XXth International Congress of History of Science (Liège, 20–26 July 1997)*, edited by Denis Buican and Denis Thieffry, 11–30. Turnhout, Belgium: Brepols, 2002.

———. "Review of H. E. Sigerist *Autobiographical Writings* and *A Bibliography of the Writings of Henry Sigerist.*" *Isis* 59 (1968): 463–64.

———. *Santorio Santorio i njegovi aparati i instrumenti.* Zagreb: Jugoslavenska akademija znanosti i umjetnosti, 1952.

———. "The Study and Teaching of the History of Medicine in Yugoslavia." *Liječnički vjesnik* 84 (1962): 5–22.

———. "Un mémoricide." In *La guerre comme maladie sociale et autres textes politiques*, 29–32. Paris: Seuil, 2001.

Grmek, Mirko D., Robert Cohen, and Guido Cimino. *On Scientific Discovery: The Erice Lectures 1977*, edited by Mirko D. Grmek, Robert S. Cohen, and Guido Cimino. Boston: Reidel, 1981.

Grmek, Mirko D., and Paul Delaunay. "L'art de guérir." In *La science moderne. De 1450 à 1800*, edited by René Taton, 155–67. 2nd ed. Paris: Presses Universitaires de France, 1969.

Grmek, Mirko D., Jacques Félician, Pierre Ginésy, Geneviève Krick, Jean-Paul Liauzu, and Jacques Saliba. "Entretien avec Mirko Grmek." *Césure* 9 (1995): 229–60.

Grmek, Mirko D., and Paul Giniewski. "Sida et relations internationales. Entretien avec Mirko D. Grmek." *Politique internationale* 50 (1991): 351–64.

Grmek, Mirko D., Marc Gjidara, and Neven Simac. *Le nettoyage ethnique. Documents historiques sur une idéologie serbe.* Paris: Le Seuil, 1993.

Grmek, Mirko D., O. Ouvry, D. Marcelli, and Y. Edel. "Le sida remis dans son histoire. Entretien avec le Pr Grmek." *Nervure* 3 (1990): 54–65.

Groeben, Christiane. "Stazione Zoologica Anton Dohrn." *eLS* (2013), https://doi.org/10.1002/9780470015902.a0024932.

Gruber, Howard E. *Darwin on Man: A Psychological Study of Scientific Creativity.* New York: Dutton, 1974.

Gruner, Charles G. *Morborum antiquitates.* Vratislavia, Bohemia: Korn, 1774.

Grunfeld, C., and K. R. Feingold, "Metabolic Disturbances and Wasting in the Acquired Immunodeficiency Syndrome." *New England Journal of Medicine* 327 (1992): 329–37.

Guerra, Francisco. "Cause of Death of the American Indians." *Nature* 326 (1987): 449–50.

———. "The Dispute over Syphilis: Europe versus America." *Clio Medica* 13 (1978): 39–61.

———. "The Earliest American Epidemic: The Influenza of 1493." *Social Science History* 12 (1988): 305–25.

———. "The Influence of Disease on Race, Logistics and Colonization in the Antilles." *Journal of Tropical Medicine and Hygiene* 69 (1966): 23–35.

———. "La invasion de America por virus." In *Maladie et maladies, histoire et conceptualisation (Mélanges en l'honneur de Mirko Grmek)*, edited by Danielle Gourevitch, 221–228. Geneva: Droz, 1992.

Gundulić, Ivan. *Dubravaka.* Dubrovnik, 1837.

Hadamard, J. *An Essay on the Psychology of Invention in the Mathematical Field.* 2nd ed. New York: Dover, 1954.

Haeser, Heinrich. *Lehrbuch der Geschichte der Medicin und der epidemischen Krankheiten.* Jena, Germany: Mauke, 1882.

Hahn, B. "Biologically Unique, SIV-like HIV–A Variants in Healthy West African Individuals." In *Cinquième Colloque des Cents-Gardes*, edited by M. Girard and L. Valette, 31–38. Lyon, France: Mérieux, 1990.

Hahn, B. H., G. M. Shaw, M. E. Taylor, R. R. Redfield, P. D. Markham, S. Z. Salahuddin, F. Wong-Staal, R. C. Gallo, E. S. Parks, and W. P. Parks. "Genetic Variation in HTVL–III/LAV over Time in Patients with AIDS or at Risk for AIDS." *Science* 232 (1986): 1548–53.

Hahn, Roger. "Berkeley's History of Science Dinner Club: A Chronicle of Fifty Years of Activity." *Isis* 90 (1999): 182–91.

———. "Sarton Medal Citation." *Isis* 83 (1992): 281–82.

Hall, G. Stanley. *Senescence: the Last Half of Life.* New York: Appleton, 1922.

Haller, Albrecht von. *Physiological Elements of the Human Body.* 1757–1766.

Halpern, D. A. "Alexander A. Bogomolets." *American Review of Soviet Medicine* 1 (1943): 173–75.

Hancock, G., and E. Carim. *AIDS: The Deadly Epidemic.* London: Gollancz, 1986.

Hanson, Norwood Russel. "The Idea of a Logic of Discovery." *Dialogue: Canadian Philosophical Review* 4 (1965): 48–61.

———. *Patterns of Discovery.* London: Cambridge University Press, 1958.

Hare, Ronald. *Pomp and Pestilence.* London: Gollancz, 1954.

Hargreaves, R. "The Saving Quality." *Practitioner* 190 (1963): 263–71.

Harvey, William. "An Anatomical Disquisition on the Motion of the Heart and Blood in Living Beings." In *The Works of William Harvey.* Translated by Robert Willis, with an introduction by Arthur C. Guyton. Philadelphia: University of Pennsylvania Press, 1989.

Haseltine, W. A., and F. Wong-Staal. "The Molecular Biology of the AIDS Virus." *Scientific American* 259 (1988): 52–62.

Hecker, Justus C. F. *Der schwarze Tod im vierzehnten Jahrhundert*. Berlin: Verlag, 1832.

Herdan, G. "The Mathematical Relation between the Number of Diseases and the Number of Patients in a Community." *Journal of the Royal Statistical Society*, Section A 120 (1957): 320–30.

Hirsch, August. *Handbuch der historisch-geographischen Pathologie*. 3 vols. Stuttgart, Germany: Enke, 1881–1886.

Hirst, Leonard Fabian. *The Conquest of Plague: A Study of the Evolution of Epidemiology*. Oxford: Clarendon, 1953.

Hoffmann, Friedrich. *La médecine raisonnée*. 9 vol. Translated by Jacques-Jean Bruhier. Paris: Briasson, 1739–1743.

Høfmann, Bjorn. "On the Triad Disease, Illness and Sickness." *Journal of Medicine and Philosophy* 27 (2002): 651–73.

Holcomb, Richmond C. *Who Gave the World Syphilis? The Haitian Myth*. New York: Froben, 1930.

Holland, J. et al. "Rapid Evolution of RNA Genomes." *Science* 215 (1982): 1577–88.

Holmes, Frederic L. *Claude Bernard and Animal Chemistry: The Emergence of a Scientist*. Cambridge, Mass.: Harvard University Press, 1974.

———. *Investigative Pathways: Patterns and Stages in the Careers of Experimental Scientists*. New Haven, Conn.: Yale University Press, 2004.

Holmes, Frederic L., Jürgen Renn, and Hans-Jörg Rheinberger. *Reworking the Bench: Research Notebooks in the History of Science*. Dordrecht, Netherlands: Kluver, 2003.

Holsendorf, B. E. "The Rat and Ratproof Construction of Buildings." Suppl. no. 131. Washington, DC: U.S. Public Health Service, 1937.

Hoyle, Fred, and Nalin C. Wickramasinghe. *Diseases from Space*. London: J. M. Dent, 1979.

Huard, Pierre. "La médecine et l'histoire." *Revue de synthèse* 37–39 (1965): 103–30.

Hudson, Ellis H. "Christopher Columbus and the History of Syphilis." *Acta Tropica* 25 (1968): 1–16.

———. *Non-Venereal Syphilis: A Sociological and Medical Study of Bejel*. London: E. S. Livingstone, 1958.

Hudson, Robert. "How Diseases Birth and Die." *Transactions of the College of Physicians of Philadelphia* 45 (1977): 18–27.

Huet, Thierry, Rémi Cheynier, Andreas Mayerhans, Georges Roelants, and Simon Wain-Hobson. "Genetic Organization of a Chimpanzee Lentivirus Related to HIV–1." *Nature* 345 (1990): 356–59.

Huet, Thierry, Marie-Christine Dazza, Françoise Brun-Vézinet, Georges E. Roelants, and Simon Wain-Hobson. "A Highly Defective HIV–1 Strain

Isolated from a Healthy Gabonese Individual Presenting an Atypical Western Blot." *AIDS* 3 (1989): 705–15.

Hufeland, Christoph W. *Die Kunst das menschliche Leben zu verlängern*. Jena, Germany: Akademische Buchhandlung, 1797.

Huisman, Frank. "The Dialectics of Understanding: On Genres and the Use of Debate in Medical History." *History and Philosophy of the Life Sciences* 27 (2005): 13–40.

Huisman, Frank, and John Harley Warner, eds. *Locating Medical History: The Stories and Their Meaning*. Baltimore, Md.: Johns Hopkins University Press, 2004.

Jauvert, Vincent. "La rumeur du KGB." *Le Nouvel Observateur*, June 11, 1992, 20–21.

Jevons, William Stanley. *The Principles of Science: A Treatise on Logic and Scientific Method*. London: Macmillan, 1874.

Jones, Colin. "The Pathocenosis Moment: Mirko Grmek, the *Annales* and the Vagaries of Longue Durée." *History and Philosophy of the Life Sciences* 27 (2005): 5–11.

Khalife, Jamal, Jean-Marie Grzych, Raymond Pierce, Jean-Claude Ameisen, Anne-Marie Schacht, Hélène Gas-Masse, André Tartar, Jean-Pierre Lecocq, and André Capron. "Immunological Crossreactivity between the Human Immunodeficiency Virus Type 1 Virion Infectivity Factor and a 170-kD Surface Antigen of Schistosoma Mansoni." *Journal of Experimental Medicine* 172 (1990): 1001–4.

Kilbourne, Edward D. *Influenza*. New York: Plenum, 1987.

Klein, Alexandre. "Quelle place pour Mirko D. Grmek, élève de Georges Canguilhem, dans l'historiographie médicale française?" In *Médecine, science, histoire. Le legs de Mirko Grmek*, edited by Pierre-Olivier Méthot. Paris: Matériologiques, forthcoming.

Klein, Marc. *Histoire des origines de la théorie cellulaire*. Paris: Hermann, 1936.

Kotovsky, D. "Alte une neue Wege in der Erfoschung des Alterns." *Sudhoffs Archiv fur Geschichte der Medizin und der Naturwissenschaften* 38 (1954): 58–70.

———. "R. Rössle und die Altersforschung." *München Medical Wochenschr* 99 (1957): 1510–11.

Krieg-Planque, Alice. *"Purification ethnique." Une formule et son histoire*. Paris: CNRS Éditions, 2003.

Kuhar, Martin. "'From an Impure Source, All Is Impure': The Rise and Fall of Andrija Štampar's Public Health Eugenics in Yugoslavia." *Social History of Medicine* 30 (2016): 92–113.

Kuhn, Thomas S. *The Structure of Scientific Revolutions*. Chicago: University of Chicago Press, 1962.

Kuhn, W. "Mögliche Beziehungen der optischen Aktivität zum Problem des Alterns." *Experientia* 11 (1955): 429–36.

———. "Optische Aktivität und Begrenztheit der Lebensdauer." *Z. Altersforsch* 1 (1939): 325–41.

Kunze, H. *Forschung und Fortschritte* 9 (1925): 25.

Laignel-Lavastine, Maxime. *Histoire générale de la médecine, de la pharmacie, de l'art dentaire et de l'art vétérinaire*. 3 vols. Paris: Albin Michel, 1934–1949.

Lamartine, Alphonse de. "Le lac de b. . . ." In *Méditations poétiques*, 46–49. Paris: Librarie grecque-latine-allemande, 1820.

Lambrichs, Louise L. "Mirko D. Grmek: Bibliographie chronologique 1946– 2000." In *La vie, les maladies et l'histoire*, by Mirko D. Grmek, 175–265. Paris: Seuil, 2001.

———. "Nettoyage ethnique: le procès." In *La guerre comme maladie sociale*, by Mirko D. Grmek, 247–57. Paris: Seuil, 2001.

———. "Note sur la préhistoire de ce volume." In *Le concept de pathocénose de Mirko Grmek. Une conceptualisation novatrice de l'histoire des maladies*, edited by Joël Coste, Bernardino Fantini, and Louise L. Lambrichs, 9–12. Geneva: Droz, 2016.

———. "Un intellectuel européen engagé." In *La vie, les maladies et l'histoire*, 83–174. Paris: Seuil, 2001.

Lampert, H. "Die kolloidchemische Seite des Alterns und ihre Bedeutung für die Entstehung und Behandlung einiger Krankheiten." *Zeitschfrift Alternsforschung* 1 (1938): 96–114.

Lange-Eichbaum, Wilhelm. *Genie, Irrsinn und Ruhm*. 5th ed. Munich, Germany: Reinhardt, 1961.

Langer, William L. "The Next Assignment." *The American Historical Review* 63 (1958): 283–304.

Laqueur, Thomas W. "Viral Cultures." *The New Republic* (July 8, 1991): 36–41.

Lattimer, G., and R. A. Osborne. *Legionnaire's Disease*. New York: Dekker, 1981.

Laurens, André Du. *Discours de la conservation de la vue, des maladies mélancoliques, des catarrhes, et de la vieillesse*. Paris: J. Mettayer, 1597.

———. "Quatrième discours auquel est traitté de la vieillesse, et comme il la faut entretenir." In *Les Oeuvres de M. André du Laurens*. Rouen, France: Raphael du petit val, 1621.

Laver, W. Graeme. *The Origin of Pandemic Influenza Viruses*. New York: Elsevier, 1983.

Learmont, Jennifer, Brett Tindall, Louise Evans, Anthony Cunningham, Philip Cunningham, John Wells, Ronald Penny, John Kaldor, and David A. Cooper. "Long-Term Symptomless HIV–1 Infection in Recipients of Blood Products from a Single Donor." *Lancet* 340 (1992): 863–67.

Le Cat, Claude-Nicolas. *Dissertation sur un homme artificiel dans lequel on verrait plusieurs phénomènes de l'homme vivant*. Académie de Rouen, 1744.

Lecomte du Noüy, Pierre. *Le temps et la vie*. Paris: Gallimard, 1936.

LeDuc, J. W. "Epidemiology of Hemorrhagic Fever Viruses." *Clinical Infectious Diseases* 2, suppl. 4 (1989): 730–35.

Leibowitch, J. "Wasting as the Ultimate Defense, or the Metchnikoff-Gandhi Macrophage." Paper presented at the Seventh Cent Gardes Meeting, Marnes-la-Coquette, France, 1992.

Lemaitre, M., Y. Henin, F. Destouesse, C. Ferrieux, L. Montagnier, and A. Blanchard. "Protective Activity of Tetracycline Analogs against the

Cyctopathic Effect of the Human Immunodeficiency Virus in CEM Cells." *Research in Virology* 141 (1990): 5–16.

———. "Role of Mycoplasma Infection in the Cytopathic Effect Induced by the Immunodeficiency Virus Type 1 in Infected Cell Lines." *Infection and Immunity* 60 (1992): 742–48.

Lenski, Richard E. "Evolution of Plague Virulence." *Nature* 334 (1988): 473–74.

Léonard, Jacques. *La médecine entre les pouvoirs et les savoirs*. Paris: Aubier Montaigne, 1981.

Lépine, R. *Le diabète sucré*. Paris: Alcan, 1909.

Loeb, Jacques. "Natural Death and the Duration of Life." *Scientific Monthly* 9 (1919): 578–85.

———. "Ueber die Ursache des natürlichen Todes." *Pflügers Archiv für die gesamte Physiologie* 124 (1908): 411–26.

Lorand, Arnold. *Das Altern, seine Ursache und Behandlung*. Leipzig, Germany: J. A. Barth, 1932.

———. "Problem of Rejuvenation." *Lancet* 1 (1931): 189–90.

———. "Quelques considérations sur les causes de la sénilité." *Compte rendus de la Société de Biologie* 57 (1904): 500–2.

Lowe, John, and F. McNulty, "Tuberculosis and Leprosy: Immunological Studies in Healthy Persons." *British Medical Journal* 2 (1953): 579–84.

Luciani, Luigi. *Human Physiology*. London: Macmillan, 1921.

Lumière, Auguste. *Sénilité et rajeunissement*. Paris: André Lesot, 1932.

———. *Théorie colloidale de la biologie et de la pathologie*. Paris: Chiron, 1922.

Lutard-Tavard, Catherine. "Être à la barre, être accusé(e)." *Socio* 3 (2014): 63–78.

MacGregor, Robert. "An Experimental Inquiry into the Comparative State of Urea in Healthy and Diseased Urine and the Seat of the Formation of Sugar in Diabetes Mellitus: An Essay." *London Medical Gazette* 20 (1837): 221–24.

Madkour, M. Monir. "Historical Aspects of Brucellosis." In *Brucellosis*, edited by M. Monir Madkour, 15–20. London: Butterworths, 1989.

Magendie, François. "Note sur la présence normale du sucre dans le sang." *Compte Rendus de l'Académie des Sciences* 23 (1846): 189.

Major, Ralph H. *Disease and Destiny*. New York: Appleton, 1936.

———. "War and Disease." *Journal of Laboratory and Clinical Medicine* 28 (1943): 661–67.

Manchester, Keith. "Leprosy: The Origin and Development of the Disease in Antiquity." In *Maladie et maladies. Histoire et conceptualisation (Mélanges en l'honneur de Mirko Grmek)*, edited by Danielle Gourevitch, 31–52. Geneva: Droz, 1992.

Mareuil, J. de, B. Brichacek, D. Salaun, J. C. Chermann, and I. Hirsch. "The Human Immunodeficiency Virus (HIV) Gag Gene Product p18 Is Responsible for Enhanced Fusogenicity and Host Range Tropism of Highly Cytopathic HIV–1–NDK Strain." *Journal of Virology* 66 (1992): 6797–6801.

Marinescu, G. "Mécanisme chimico-colloidal de la sénilité et le problème de la mort naturelle." *Revue des sciences* 1 (1914): 673–79.

———. "Nouvelle contribution à l'étude du mécanisme de la vieillesse." *Bulletin de l'Académie de Médecine* 111 (1934): 761–72.

Martini, Erich. *Die Wege der Seuchen*. Stuttgart, Germany: Enke, 1954.

Martini, Gustav A., and Rudolf Siegert. *Marburg Virus Disease*. Berlin: Springer, 1971.

Mattern, Susan P. *The Prince of Medicine: Galen in the Roman Empire*. Oxford: Oxford University Press, 2013.

Maulitz, Russel C. "Reflections on Yellow Fever and AIDS." In *Maladie et maladies, histoire et conceptualisation (Mélanges en l'honneur de Mirko Grmek)*, edited by Danielle Gourevitch, 399–414. Geneva: Droz, 1992.

May, Jacques M. *Ecology of Human Disease*. New York: M. D. Publications, 1959.

———. *Studies in Disease Ecology*. New York: Hafner, 1961.

McDade, Joseph E., Charles C. Shepard, David W. Fraser, Theodore R. Tsai, Martha A. Redus, Walter R. Dowdle, and the laboratory investigation team. "Legionnaire's Disease: Isolation of a Bacterium and Demonstration of Its Role in Other Respiratory Diseases." *New England Journal of Medicine* 297 (1977): 1197–1203.

McKeown, Thomas. *The Origins of Human Diseases*. London: Blackwell, 1988.

McNeill, William H. *Plagues and People*. New York: Doubleday, 1976.

McNicol, L. A., and R. N. Doetsch. "A Hypothesis Accounting for the Origins of Pandemic Cholera: A Retrograde Analysis." *Perspectives in Biology and Medicine* 26 (1983): 547–52.

Medvedev, Zhores A. "AIDS Virus Infection: A Soviet View of Its Origin." *Journal of the Royal Society of Medicine* 79 (1986): 494–95.

———. "Rol nervnoi sistemy v procese starenija organizma." *Prioda* 3 (1953): 101–4.

———. "Teorija prof. A. V. Nagornogo o stareniji organizma." *Fiziolgicheskii zhurnal SSSR* 38 (1952): 523–29.

Metchnikoff, Élie. "Études biologiques sur la vieillesse." *Annales de l'Institut Pasteur* 16 (1901–1902): 864–79, 913–17.

———. *Études sur la nature humaine: Essai de philosophie optimiste*. Paris: Masson, 1903.

———. *The Prolongation of Life: Optimistic Studies*. London: Heinmann, 1907.

Méthot, Pierre-Olivier. "Le concept de pathocénose chez Mirko Grmek: une réflexion évolutionniste sur l'écologie des maladies?" In *Le concept de pathocénose de Mirko Grmek. Une conceptualisation novatrice de l'histoire des maladies*, edited by Joël Coste, Bernardino Fantini, and Louise Lambrichs, 93–117. Geneva: Droz, 2016.

———. "De la pathocénose aux maladies émergentes: production, circulation et transformation conceptuelles." In *Médecine, science, histoire. Le legs de Mirko Grmek*, edited by Pierre-Olivier Méthot (Paris: Matériologiques, forthcoming).

———. "Mirko Grmek et l'histoire de la médecine et des sciences au XXe siècle." In *Médecine, science, histoire. Le legs de Mirko Grmek*, edited by Pierre-Olivier Méthot. Paris: Matériologiques (forthcoming).

———. "Why Do Parasites Harm Their Host? On the Origin and Legacy of Theobald Smith's 'Law of Declining Virulence'—1900–1980." *History and Philosophy of the Life Sciences* 34 (2012): 361–401.

Méthot, Pierre-Olivier, ed. *Médecine, science, histoire. Le legs de Mirko Grmek.* Paris: Matériologiques, forthcoming.

Méthot, Pierre-Olivier, and Bernardino Fantini, "Medicine and Ecology: Historical and Critical Perspectives on the Concept of 'Emerging Disease.'" *Archives Internationales d'histoire des sciences* 64 (2014): 213–30.

Mettrie, Julien Offroy de La. *Man a Machine.* Translated by Gertrude C. Bussey, revised by M. W. Calkins. Chicago: Open Court, 1912.

Meyer, Jean. "Une enquête de l'Académie de médecine sur les épidémies (1774–1794)." *Annales: Économies, Sociétés, Civilisations* 21 (1966): 729–49.

Meyerson, Émile. *Du cheminement de la pensée.* Paris: Alcan, 1931.

Miescher, K. "Zur Frage der Alternsforschung." *Experientia* 11 (1955): 417–29.

Minot, C. S. "On the Nature and Cause of Old Age." *Harvey Lecture* 1 (1906): 230–50.

———. *The Problem of Age, Growth and Death: A Study on Cytomorphosis.* London: Murray, 1908.

Möbius, Karl August. *Die Auster und dieAusternwirthschaft.* Berlin: Wiegandt, Hempel, and Parey, 1877.

Mollaret, Henri H. "Interprétation socio-écologique de l'apparition de maladies réellement nouvelles." In *Sida, épidémies et sociétés*, 108–14. Lyon, France: Fondation Marcel Mérieux, 1987.

———. "A Personal View of the History of the Genus *Yersinia*." *Contributions to Microbiology and Immunology* 9 (1987): 1–13.

Moller-Christensen, V. "Evidence of Tuberculosis, Leprosy and Syphilis in Antiquity and the Middle Ages." In *Proceedings of the 19th International Congress of the History of Medicine*, 229–37. New York: Karger, 1966.

Monath, T. P. "Lassa Fever: Review of Its Epidemiology and Epizootiology." *Bulletin of the World Health Organization* 52 (1975): 577–92.

Montagnier, L. "Le rétrovirus de l'immunodéficience chez l'homme et les primates." *Bulletin de l'Académie Nationale de Médecine* 173 (1989): 158–90.

———. "Origin and Evolution of HIVs and Their Role in AIDS Pathogenesis." *Journal of Acquired Immunological Deficiency Syndrome* 1 (1988): 517–20.

Morin, Edgar. *Le paradigme perdu: la nature humaine.* Paris: Le Seuil, 1973.

Morse, Stephen S., ed. *Emerging Viruses.* Oxford: Oxford University Press, 1993.

———. "Emerging Viruses: Defining the Rules for Viral Traffic." *Perspectives in Biology and Medicine* 34 (1991): 387–409.

Morse, Stephen S., and Ann Schluederberg. "Emerging Viruses: The Evolution of Viruses and Viral Diseases." *Journal of Infectious Diseases* 162 (1990): 1–7.

Moulin, Anne-Marie. "La métaphore vaccine. De l'inoculation à la vaccination." *History and Philosophy of the Life Sciences* 14 (1992): 271–97.

Mugler, Charles. "Démocrite et le danger de l'irradiation cosmique." *Revue d'histoire des sciences* 20 (1967): 221–28.

Mühlmann, M. S. "L'état actuel de la question du vieillissement." *Scientia*, *Milano* 60 (1936): 327–38.

———. *Uchenie o roste, starosti I smerti*. Baku, 1926.

———. *Ueber die Ursache des Alterns*. Wiesbaden, Germany: Bergmann, 1900.

Müller-Wille, Staffan. "History of Science and Medicine." In *The Oxford Handbook of the History of Medicine*, edited by Mark Jackson, 469–83. Oxford: Oxford University Press, 2011.

Münster, L. "Il primo trattato pratico compiuto sui problem della vecchiaia." *Rivista di Gerontologia e Geriatria* 1 (1951): 38–54.

Murray, Ian. "Paulesco and the Isolation of Insulin." *Journal of the History of Medicine and Allied Sciences* 26 (1971): 150–57.

Myers, G., K. MacInnes, and B. Korber. "The Emergence of Simian/Human Immunodeficiency Viruses." *AIDS Research in Human Retroviruses* 8 (1992): 373–86.

Nagel, Ernest. *The Structure of Science*. New York: Harcourt, 1961.

Nagornyi, A. V. *Problema starenija i dolgoletija*. Kharkiv, Ukraine: Kharkov State University Press, 1940.

———. *Starenije i prodlenije zhizni*. Moscow, 1950.

Nascher, Ignatz L. "Geriatrics." *New York Medical Journal* 90 (1909): 358–59.

Needham, Joseph. *Man a Machine*. London: Kegan Paul, 1927.

Neuberger, Max. *Geschichte der Medizin*. Vol. 1. Stuttgart, Germany: Verlag, 1906.

Nicolle, Charles. *Destin des maladies infectieuses*. Paris: Alcan, 1933.

Nohl, J. *Der Schwarze Tod. Ein Chronik der Pest 1348 bis 1720*. Potsdam, Germany: Kiepenheuer, 1924.

Nouchi, F. "La transparence de l'information en URSS doit s'appliquer au sida." *Le Monde*, November 6, 1987.

Nowak, M. "HIV Mutation Rate." *Nature* 347 (1990): 347.

O'Connor, W. "Herd Immunity and the HIV Epidemic." *Preventive Medicine* 20 (1991): 329–42.

Olby, Robert. *The Path to the Double Helix*. London: Macmillan, 1974.

Olmsted, James M. D. *Claude Bernard, Physiologist*. New York: Harper, 1938.

Olmsted, James M. D., and E. Harris Olmsted. *Claude Bernard and the Experimental Method in Medicine*. New York: Schuman, 1952.

O'Malley, Charles, D., ed. *The History of Medical Education: An International Symposium Held February 5–9, 1968*. Los Angeles: University of California Press, 1970.

Oppenheimer, Gerald M. "Review of Mirko D. Grmek. *History of AIDS: Emergence and Origin of a Modern Pandemic*." *Isis* 83 (1992): 693–94.

Oriel, J. D., and Aidan Cockburn. "Syphilis: Where Did It Come From?" *Paleopathology Newsletter* 6 (1974): 9–12.

Pagel, Walter. *New Light on William Harvey*. Basel, Switzerland: Karger, 1976.

Panum, Peter Ludwig. "Observations Made during the Epidemic of Measles on the Faroe Islands in the Year 1846." In *Medical Classics*, edited by C. Kelly Emerson, vol. 3, 829–886. New York: Williams and Wilkins, 1939.

Parhon, Constantin I. *Bătrâneţea şi tratamentul ei problema reîntineririi*. Bucharest: Editura Academiei Republicii Populare Române, 1948.

———. *Biologia vîrstelor. Cercetari clinice si exeperimentale*. Bucharest: Editura Academiei Republicii Populare Române, 1955.

Pascal, Blaise. *Pascal's Pensées*. Translated by W. F. Trotter. With an introduction by T. S. Eliot. New York: Dutton, 1958.

Pasteur, Louis, Charles E. Chamberland, and Émile Roux. "De l'atténuation des virus et de leur retour à la virulence." *Compte Rendus de l'Académie des Sciences* 92 (1881): 429–35.

Pavlov, Ivan P. *Conditioned Reflexes and Psychiatry: Lectures on Conditioned Reflexes*. Vols. 1–2. Translated by W. H. Gantt. New York: International Publishers, 1927.

———. *Polnoe sobranie sočinenij*. 2nd ed. Moscow: Akademii Nauk SSSR, 1951–1952.

———. *Sämtliche Werke*. Berlin: Akademie, 1954–1955.

Pavlovsky, Evgeny, N. *Natural Nidality of Transmissible Diseases: With Special Reference to the Landscape Epidemiology of Zooanthroponoses*. Translated by Frederick K. Plous Jr. Urbana: University of Illinois Press, 1966.

Pazzini, Adalberto, and Aroldo Baffoni. *Storia delle malattie*. Rome: Clinica nuova, 1950.

Peeters, M. et al. "Isolation and Partial Characterization of an HIV-Related Virus Occurring Naturally in Chimpanzees in Gabon." *AIDS* 3 (1989): 625–30.

Pépin, Jacques. *The Origins of AIDS*. Cambridge: Cambridge University Press, 2011.

Pereira, Gómez. *Antoniana Margarita: opus nempe physicis, medicis ac theologis non minus vtile quam necessarium. per Gometium Pereiram, medicum Methinæ Duelli, quae Hispanorum lingua Medina de el Campo apellatur, nunc primum in lucem æditum*. Medina del Campo, Spain, 1554.

Perrenoud, A. "Contributions à l'histoire cyclique des maladies. Deux cents ans de variole à Genève (1580–1810)." In *Mensch und Gesundheit in der Geschichte*, edited by Arthur E. Imhof, 175–78. Husum, Germany: Matthiesen, 1980.

Peter, Jean-Pierre. "Une enquête de la Société Royale de médecine: malades et maladies à la fin du XVIIIe siècle." *Annales: Économies, Sociétés, Civilisations* 22 (1967): 711–51.

Phillippe, A. *Histoire de la peste noire (1348–1350) d'après des documents inédits*. Paris, 1853.

Piaget, Jean. *Biology and Knowledge: An Essay on the Relations between Organic Regulations and Cognitive Processes*. Chicago: University of Chicago Press, 1971.

Piel, Gerard, Dennis Flanagan, Leon Svirsky, George A. Boehm, Robert Emmett Ginna, Jean Le Corbeiller, James R. Newman, E. P. Rosenbaum, and James Grunbaum, eds. *The Physics and Chemistry of Life*. New York: Simon and Schuster, 1955.

Plato. *Timaeus*. Translated by Benjamin Jowett. New York: Macmillan, 1987.

Pliny the Elder. *The Natural History*. Translated by John Bostock. London: Taylor and Francis, 1855.

Plutarch. *Propos de table. Oeuvres morales de Plutarque*. Translated by Ricard. Paris: Didier, 1844.

Politzer, R. *La peste*. Geneva: WHO, 1954.

Pollack, Michaël. "Comptes rendus: *Histoire du sida. Début et origine d'une pandémie actuelle*." *Annales E.S.C.* 44 (1989): 1521–23.

Popper, Karl. *The Logic of Scientific Discovery*. New York: Routledge, 2002.

Porter, Roy. "The Patient's View: Doing History from Below." *Theory and Society* 14 (1985): 175–98.

———. "Review of *Histoire du sida: Début et origine d'une pandémie actuelle*." *Medical History* 34 (1990): 458–59.

Pratt, Joseph H. "A Reappraisal of Researches Leading to the Discovery of Insulin." *Journal of the History of Medicine and Allied Sciences* 9 (1954): 281–89.

Preston, Frank W. "The Commonness, and Rarity, of Species." *Ecology* 29 (1948): 254–83.

Pütter, A. "Lebensdauer und Alternsfaktor." *Zeitschrift für allgemeine Physiologie* 19 (1921): 9–36.

Rabello, F. E. "Les origines de la syphilis." *Nouvelle presse médicale* 2 (1973): 1376–80.

Ratcliff, Marc J. "Journée d'étude 'Théorie et méthode dans les sciences de la vie,' Geneva, 26 January 1996." *Gesnerus* 53 (1996): 248–50.

Ray, John. *The Wisdom of God Manifested in the Works of the Creation*. London: R. Harbin, 1691.

Razzell, Peter. *Edward Jenner's Cowpox Vaccine: The History of a Medical Myth*. Lewes, Del.: Caliban, 1977.

Redondi, Pietro. "Ernest Coumet et l'histoire de l'histoire des sciences." *Revue de synthèse* 4 (2001): 291–96.

Reichinstein, D. *Das Problem des Alterns und die Chemie der Lebensvorgänge*. 2nd ed. Zurich: Akerets Erben, 1940.

Reisel, Salomon. "Statua humana circulatoria." In *Miscellana Curiosa, Decuria I, Annus IX/X für 1678/9, Observatio I*. Stuttgart, Germany: Sipho Würrtembergicus, 1690.

Renouard, Y. "Conséquences et intérêt démographique de la peste noire de 1348." *Population* 3 (1948): 459–66.

Revel, Jean-François. *La connaissance inutile*. Paris: Grasset, 1988.

Rey, Abel. *L'énergétique et le mécanisme. Au point de vue des conditions de la connaissance*. Paris: Félix Alcan, 1908.

Rheinberger, Hans-Jörg. "History of Science and the Practice of Experiment." *History and Philosophy of the Life Sciences* 23 (2001): 51–63.

———. *On Historicizing Epistemology*. Translated by David Fernbach. Stanford, Calif.: Stanford University Press, 2010.

———. *Toward a History of Epistemic Things: Synthesizing Proteins in the Test Tube*. Stanford: Stanford University Press, 1997.

Risse, Guenter B. "Review of Mirko D. Grmek. *History of AIDS: Emergence and Origin of a Modern Pandemic*." *Bulletin of the History of Medicine* 65 (1991): 604–6.

Robert, Fernand. "Un événement dans les études grecques: Le nouveau livre de Mirko D. Grmek." *Bulletin de l'Association Guillaume Budé* 2 (1984): 213–20.

Roberts, R. S. "A Consideration of the Nature of the English Sweating Sickness." *Medical History* 9 (1965): 385–89.

———. "Epidemics and Social History: An Essay Review." *Medical History* 12 (1968): 305–16.

Robertson, T. B. *The Chemical Basis of Growth and Senescence*. Philadelphia, Penn.: Lippincott, 1923.

Rocasolano, A. de Gregorio. "Physikalisch-chemische Hypothese über das Altern." *Kolloidchemie Beihefte* 19 (1924): 441–76.

Roger, Jacques. "Pour une histoire historienne des sciences." In *Pour une histoire des sciences à part entière*, edited by Claude Blanckaert, 45–73. Paris: Albin Michel, 1995.

Rolleston, John D. *The History of the Acute Exanthemata*. London: Heinemann, 1937.

Rosen, Leon. "Dengue in Greece in 1927 and 1928 and the Pathogenesis of Dengue Hemorrhagic Fever: New Data and a Different Conclusion." *The American Journal of Tropical Medicine* 35 (1986): 642–53.

Rosenberg, Charles E. "Commentary." In *A Celebration of Medical History: The Fiftieth Anniversary of the Johns Hopkins Institute of the History of Medicine and the Welch Medical Library*, edited by Lloyd G. Stevenson, 32–36. Baltimore, Md.: Johns Hopkins University Press, 1982.

———. "Disease in History: Frames and Framers." *The Milbank Quarterly* 67, no. S1 (1989): 1–15.

Rosqvist, Roland, Mikael Skurnik, and Hans Wolf-Watz. "Increased Virulence of *Yersinia* Pseudo-Tuberculosis by Two Independent Mutations." *Nature* 334 (1988): 522–25.

Rössle, Robert. *Wachstum und Altern*. Munich, Germany: J. F. Bergmann, 1923.

Rothschild, B. M., and W. Turnbull. "Treponemal Infection in a Pleistocene Bear." *Nature* 329 (1987): 61–62.

Rubner, Max. *Das Problem der Lebensdauer une seine Beziehungen zu Wachstum und Ernährung*. Berlin: Oldenbourg, 1908.

Rübsamen-Waigman, H., and U. Dietrich. "Die Ahnen des AIDS-Virus." *Bild der Wissenschaft* 3 (1991): 92–96.

Rupke, Nicolaas A. "*Bathybius Hackelii* and the Psychology of Scientific Discovery." *Studies in History and Philosophy of Science Part A* 7 (1976): 53–62.

Russell, Josiah C. "Effects of Pestilence and Plague, 1315–1385." *Comparative Studies in Society and History* 8 (1966): 464–73.

Růžička, Vladislav. "Beitrage zum Studium der Protoplasmahysteresis und der hysteretischen Vorgänge, I. Die Protoplasmahysteresis als Entropieerscheinung." *Archiv für Mikroskopiche Anatomie Entwicklung Mechanik* 101 (1924): 459–82.

Sabatier, René. *Sida: l'épidémie raciste*. Paris: L'Harmattan, 1989.

Sabin, Albert B. "Nature of Inherited Resistance to Viruses." *Proceedings of the National Academy of Science* 38 (1952): 540–46.

Saharov, G. P. *La Lutte contre la vieillesse selon Metchnikoff*. Moscow, 1938.

Sallares, Robert. *The Ecology of the Ancient Greek World*. Ithaca, N.Y.: Cornell University Press, 1991.

———. "Pathocenosis: Ancient and Modern." *History and Philosophy of the Life Sciences* 27 (2005): 201–20.

Salomon Bayet, Claire. "L'histoire des sciences et des techniques." In *L'histoire et le métier d'historien en France 1945–1995*, edited by François Bédarida, 379–392. Paris: Maison des sciences de l'homme, 1995.

Santorio, Santorio. *Medicina Statica: Being the Aphorisms of Sanctorius*. Translated by John Quincy. London: W. and Newton in Little Britain, A. Bell at the Cross-Keys in Cornhill, W. Taylor at the Ship in Paternoster-Row, and J. Osborn at the Oxford-Arms in Lombard-Street, 1728.

Scheffler, Israel. *The Anatomy of Inquiry: Philosophical Studies in the Theory of Science*. New York: Knopf, 1963.

Schlomka, G. "Über Ziele und Wege klinischer Alternsforschung." In *Festschift M. Bürger*, 417–44. Leipzig, Germany: Wiss Z., 1955.

Scott, H. Harold. *A History of Tropical Medicine*. London: Edward Arnold, 1939.

Seale, John. "AIDS Virus Infection: Prognosis and Transmission." *Journal of the Royal Society of Medicine* 78 (1985): 613–15.

———. "Artificial HIV?" *Nature* 335 (1988): 391.

———. "Crossing the Species Barrier: Viruses and the Origin of AIDS in Perspective." *Journal of the Royal Society of Medicine* 82 (1989): 519–23.

———. "Origins of the AIDS Viruses, HIV–1 and HIV–2: Fact or Fiction? Discussion Paper." *Journal of the Royal Society of Medicine* 81 (1988): 537–39.

Seale, John, and Zhores A. Medvedev. "Origin and Transmission of AIDS: Multi-Use Hypodermics and the Threat to the Soviet Union: Discussion Paper." *Journal of the Royal Society of Medicine* 80 (1987): 301–4.

Segal, Jakob, Lilli Segal, and Ronald Dehmlow, *AIDS: Its Nature and Origin*. Bertrand Russell Peace Foundation, Australian Branch, 1986.

Seneca. *Naturales Quaestiones*. Translated by Thomas H. Corcoran. London: Heinmann, 1971.

Shaw, Brent D. "Grmek's Pathological Vision." *Social History of Medicine* 4 (1991): 329–34.

Shope, Richard E. "Influenza: History, Epidemiology, and Speculation." *Public Health Reports* 73 (1988): 165–78.

Sigerist, Henry E. "Der Aussatz auf den Hawaiischen Inseln." *Verhandlungen der Schweizer. Naturforschenden Gesellschaft* (1932): 452–53.

———. *A History of Medicine*. Vol. 1, *Primitive and Archaic Medicine*. New York: Oxford University Press, 1951.

———. *Landmarks in the History of Hygiene.* London: Oxford University Press, 1956.

———. *Man and Medicine: An Introduction to Medical Knowledge.* Translated by Margaret Galt Boise, with an introduction by William H. Welch. New York: Norton, 1932.

Soler, Léna, Sjoerd Zwart, Michael Lynch, and Vincent Israel Jost. *Science after the Practice Turn in the Philosophy, History, and Social Studies of Science.* New York: Routledge, 2014.

Southey, Robert. *Letters Written during a Short Residence in Spain and Portugal.* 2nd ed. Bristol, England: Bulgin and Rosen, 1799.

Soviet Influence Activities: A Report on Active Measures and Propaganda, 1986–1987. Washington, DC: United States Department of State, August 1987.

Štampar, Andrija. *Serving the Cause of Public Health: Selected Papers from Andrija Štampar,* edited by Mirko D. Grmek. Zagreb: School of Public Health, 1966.

———. "Some Comments on the Law for the Protection of National Health." In *The History of East-Central European Eugenics, 1900–1945,* edited by Marius Turda, 440–42. London: Bloomsbury, 2015.

Stearn, Esther W., and Allen E. Stearn. *The Effect of Smallpox on the Destiny of the Amerindians.* Boston: Bruce Humphries, 1945.

Stendhal. *On Love.* Translated by Philip Sidney Woolf and Cecil N. Sidney Woolf. New York: Brentano, 1915.

Stensen, Niels. "Discourse of the Anatomy of the Brain." In *Nicolas Steno: Biography and Original Papers of a 17th Century Scientist,* edited by Troels Kardel and Paul Maquet. Heidelberg, Germany: Springer, 2013.

Stevenson, Lloyd G. "New Diseases in the Seventeenth Century." *Bulletin of the History of Medicine* 39 (1965): 1–21.

Sticker, Georg. *Abhandlungen aus der Seuchengeschichte und Seuchenlehre.* 3 vols. Giessen, Germany: Töpelmann, 1908–1912.

Stol, M. "Leprosy: New Light from Greek and Babylonian Sources." *Jaarbericht van het Vooraziatische-Egyptisch genootshchap.* Ex Oriente lux 30 (1988): 22–31.

Sullivan, William. "On the Progress of General, Physiological, and Pathological Chemistry for the Years 1846–1847." *The Dublin Quarterly Journal of Medical Science* 13 (1849): 201–42.

Taylor-Robinson, D. "Are Mycoplasmas Involved in the Pathogenesis of AIDS?" In *Huitième Colloque des Cent-Gardes,* edited by M. Girard and L. Valette, 11–16. Lyon, France: Mérieux, 1993.

Tenner, Edward. "Revenge Theory." *Harvard Magazine* (March–April 1991): 27–30.

Thom, René. *Modèles mathématiques de la morphogenèse. Recueil de textes sur la théorie des catastrophes et ses applications.* Paris: Union générale d'éditions, 1974.

Thomas, F. M. "Pavlov's Work on Higher Nervous Activity and Its Development in the USSR." *Nature* 154 (1944): 385–88.

Thompson, D'Arcy W. *On Aristotle as a Biologist, with a Prooemion on Herbert Spencer.* Oxford: Clarendon, 1913.

Thomson, Thomas. *Chemistry of Animal Bodies*. Edinburgh: Adam and Charles Black, 1843.

Thrupp, Sylvia L. "Plague Effects in Medieval Europe." *Comparative Studies in Society and History* 8 (1966): 474–83.

Thucydides. *History of the Peloponnesian War*. Books 1–2. Translated by Charles Forster Smith. London: William Heinemann, 1956.

Tiedemann, F., and L. Gmelin. *Die Verdauung nach Versuchen*. Vol. 1. Leipzig, Germany: Groos, 1826.

Townsend, F. M. "Changes in Brain with Age." *Journal of Gerontology* 1 (1946): 401–2.

Turda, Marius, ed. *The History of East-Central European Eugenics, 1900–1945*. London: Bloomsbury, 2015.

The USSR's AIDS Disinformation Campaign. Washington, DC: Department of State, 1987.

van der Waerden, B. L. "Mendel's Experiments." *Centaurus* 12 (1968): 275–88.

Varmus, Harold E. "Naming the AIDS Virus." In *The Meaning of AIDS*, edited by Eric T. Juengst and Barbara A. Koenig, 3–11. New York: Praeger, 1989.

Vartanian, Aram. *La Mettrie's "L'Homme machine." A Study in the Origins of an Idea*. Critical edition with an introductory monography and notes. Princeton, N.J.: Princeton University Press, 1960.

Vesalius, Andreas. *On the Fabric of the Human Body*. Translated by W. F. Richardson in collaboration with J. B. Carman. San Francisco: Norman, 1998.

Vogt, C., and O. Vogt. "Ageing of Nerve Cells." *Nature* 158 (1946): 304.

Vučak, Ivica. "In Memoriam Mirko Dražen Grmek (1924–2000)." *Croatian Medical Journal* 41 (2000): 213–17.

Wain-Hobson, S., and J. P. Vartanian. "Sida; suivre la variation du virus." *La Recherche* 23 (1992): 1469–71.

Watson, A. J. "Origin of *Encephalitis Lethargica*." *China Medical Journal* 42 (1928): 427–32.

Watson, James D. *The Double Helix: A Personal Account of the Discovery of the Structure of DNA*. London: Weidenfeld and Nicholson, 1968.

Weiling, Franz. "Neue ergebnisse zur statistischen Vorgeschichte der Mendelschen Versuche." *Biometrics* 27 (1971): 709–19.

Weismann, August. *Ueber die Dauer des Lebens; ein Vortrag*. Jena, Germany: G. Fischer, 1882.

———. *Ueber Leben und Tod*. Jena, Germany: G. Fischer, 1884.

Wells, Herbert George, Julian Huxley, and George Philip Wells. *The Science of Life*. New York: The Literary Guild, 1929.

Wertheimer, Max. *Productive Thinking*. New York: Harper, 1945.

Williams, Carrington B. *Patterns in the Balance of Nature and Related Problems in Quantitative Ecology*. New York: Academic Press, 1964.

Williams, L. Pearce. "Normal Science, Scientific Revolutions and the History of Science." In *Criticism and the Growth of Knowledge*, edited by I. Lakatos and A. Musgrave, 49–50. Cambridge: Cambridge University Press, 1970.

Willis, Thomas. *Opera Omnia*. Coloniae, 1694.

Wilson, Adrian. "On the History of Disease-Concepts: The Case of Pleurisy." *History of Science* 38 (2000): 271–319.

Wilson, D. Wright. "Claude Bernard." *Popular Scientific Monthly* 84 (1917): 567–78.

Wylie, John A. H., and Leslie H. Collier, "The English Sweating Sickness (*Sudor Anglicus*): A Reappraisal." *Journal of the History of Medicine and Allied Sciences* 36 (1981): 422–45.

Yanagisawa, Ken. "The Effect of BCG Vaccination upon Occurrence of Leprosy in Nursery Children." *International Journal of Leprosy* 26 (1958): 325–27.

Young, F. G. "Claude Bernard and the Theory of the Glycogenic Function of the Liver." *Annals of Science* 2 (1937): 47–83.

Zapevalov, V. "Panic in the West, or What Is Behind the Sensation about AIDS." [In Russian.] *Literaturnaya Gazeta*, October 30, 1985.

Zeiss, Heinz. *Elias Metchnikoff, Leben und Werk*. Jena, Germany: G. Fischer, 1932.

Zeman, F. D. "Life's Later Years: Studies in the Medical History of Old Age." *Journal of Mount Sinai Hospital* 12 (1945–1946): 890.

Zerbi, Gabrielle. *Gerontocomia, scilicet de senium cura atque victu*. Rome: Prologus, 1489.

Zhdanov, Victor. *Sovteskaya Kultura*, December 5, 1985.

Zinsser, Hans. *Rats, Lice and History*. Boston: Little, Brown, 1935.

Župančič, Andrej O. *Uvod u opstu patofiziologiju coveka*. Belgrade, 1952.

Zylberman, Patrick. "Fewer Parallels than Antitheses: René Sand and Andrija Štampar on Social Medicine, 1919–1951." *Social History of Medicine* 17 (2004): 77–92.

———. *Tempêtes microbiennes. Essai sur la politique de sécurité sanitaire dans le monde transatlantique*. Paris: Gallimard, 2013.

normal science, 139, 212n5. *See also* extraordinary science
nosological reality. *See* retrospective diagnosis
Nouchi, F., 198n28
Nowak, M., 200

objectivity. *See* history of science; knowledge
observation, theory, and hypotheses, 141–44, 146
O'Connor, W., 210n57
Olby, Robert, 129, 212n9
Olmsted, James M. D., 203n9, 204n42
O'Malley, Charles D., 9, 179nn66,67,71
ontological, 1
Oppenheimer, Gerald, 16, 184n131
Optimistic Studies, 92, 206
Orbeli, L. A., 97
organism and machine analogy, 91–92, 104–5, 108, 111, 114, 116, 120, 209n2
Oriel, J. D., 194n26
Ouvry, D., 182n107

Padjen, Ante, 169
Pagel, Walter, 127, 212n6
paleopathology, 14–15, 183n118, 184n120. *See also* retrospective diagnosis
pandemic, 2, 16, 19–20, 24, 36, 41, 51, 53, 63, 65–66, 172n7, 184n131; AIDS and human behavior, 185
Panum, Peter Ludwig, 194n33
Paracelsus, Theophrastus, 92, 116
paradigm, 11, 21, 75, 89, 131, 139; paradigm shift, 22; paradigmatic reversal and crises, 138–39
parasites, 20, 53, 64
Parhon, Constantin, I., 98, 103, 208n40
Parmenides of Elea, 88
Pascal, Blaise, 187n3
Pasteur, Louis, 12, 69, 92, 137, 150, 181n95, 200n55
pathocenosis, vii, x, 3, 15, 34, 40, 201n58; and AIDS, vii, x, 4, 16–17, 53, 67, 70, 184n128; and biohistory, 3; definition of, 18–19, 33, 187n154; and ecological theory, 8, 33, 38, 184n127, 191n43; and emerging infections, 4, 16–17, 19, 53;

English translation of, xiv; dynamics of pathocenosis, x, 33, 35, 40, 70; early mention of, 186n154; as an intellectual tool, 19; mathematical models of, 38, 68, 70, 191n43; and natural history of disease, 8; and parasitology, 8, 19, 178n55; quantitative study of, 38–39, 191n43; reception of, 15, 178n55, 184nn126–27; rupture in the pathocenotic equilibrium, 20, 40, 70
pathogenetic point of view, 34
pathogenicity. *See* virulence
pathography, 8, 32
pathological realities, v, xiii, xv, xvii, 2, 15, 24
Pavić, Milan, 177
Pavlov, Ivan P., 97, 107, 207n34, 210n9
Pavlovsky, Evgeny N., 178n55
Pazzini, Adalberto, 32, 189n15
Peeters, M., 199n37
Pelouze, Theophile-Jules, 82
Pentagon, 59, 61, 63
Pépin, Jacques, 199n43
Pereira, Gómez, 107, 209n6
Perrenoud, A., 196n16
Petancius, Félix, 162
Peter, Jean-Pierre, 186n154, 191n49
Pflüger, Eduard, 134
Philippe, A., 190n33
philology, 11, 15, 185n137; philologist, 2, 15
philosophy of science, x, 11–13, 20, 167–68, 174n18, 177n52, 181n88, 185n137
phthisis. *See* tuberculosis
physiology, 8, 24–26, 119, 172n7; animal physiology, 82; as animated anatomy, 112, 210n25; in Baglivi, 162; in Descartes, 113–14; Galen's anatomo-physiology, 106; in Galileo, 109, 113; in Harvey, 112, 138, 147; human physiology, 205n5, 206n7; in Huxley, 120; and infinitesimal calculus, 113
Piaget, Jean, 21, 140–41, 182n102, 213n32
Piel, Gerard, 212n49
pillage of the heritage, 158
piqure sucrée, 129
plague, x, 3, 32, 34–37, 41, 43, 46, 51, 163, 216n9; of Athens, 32, 43; bovine,

forms of living

Stefanos Geroulanos and Todd Meyers, *series editors*

Georges Canguilhem, *Knowledge of Life*. Translated by Stefanos Geroulanos and Daniela Ginsburg, Introduction by Paola Marrati and Todd Meyers.

Henri Atlan, *Selected Writings: On Self-Organization, Philosophy, Bioethics, and Judaism*. Edited and with an Introduction by Stefanos Geroulanos and Todd Meyers.

Catherine Malabou, *The New Wounded: From Neurosis to Brain Damage*. Translated by Steven Miller.

François Delaporte, *Chagas Disease: History of a Continent's Scourge*. Translated by Arthur Goldhammer, Foreword by Todd Meyers.

Jonathan Strauss, *Human Remains: Medicine, Death, and Desire in Nineteenth-Century Paris*.

Georges Canguilhem, *Writings on Medicine*. Translated and with an Introduction by Stefanos Geroulanos and Todd Meyers.

François Delaporte, *Figures of Medicine: Blood, Face Transplants, Parasites*. Translated by Nils F. Schott, Foreword by Christopher Lawrence.

Juan Manuel Garrido, *On Time, Being, and Hunger: Challenging the Traditional Way of Thinking Life*.

Pamela Reynolds, *War in Worcester: Youth and the Apartheid State*.

Vanessa Lemm and Miguel Vatter, eds., *The Government of Life: Foucault, Biopolitics, and Neoliberalism*.

Henning Schmidgen, *The Helmholtz Curves: Tracing Lost Time*. Translated by Nils F. Schott.

Henning Schmidgen, *Bruno Latour in Pieces: An Intellectual Biography*. Translated by Gloria Custance.

Veena Das, *Affliction: Health, Disease, Poverty*.

Kathleen Frederickson, *The Ploy of Instinct: Victorian Sciences of Nature and Sexuality in Liberal Governance*.

Roma Chatterji, ed., *Wording the World: Veena Das and Scenes of Inheritance*.

Jean-Luc Nancy and Aurélien Barrau, *What's These Worlds Coming To?* Translated by Travis Holloway and Flor Méchain. Foreword by David Pettigrew.

Anthony Stavrianakis, Gaymon Bennett, and Lyle Fearnley, eds., *Science, Reason, Modernity: Readings for an Anthropology of the Contemporary*.

Richard Baxstrom and Todd Meyers, *Realizing the Witch: Science, Cinema, and the Mastery of the Invisible*.

Hervé Guibert, *Cytomegalovirus: A Hospitalization Diary*. Introduction by David Caron, Afterword by Todd Meyers, Translated by Clara Orban.

Leif Weatherby, *Transplanting the Metaphysical Organ: German Romanticism between Leibniz and Marx*.

Fernando Vidal and Francisco Ortega, *Being Brains: Making the Cerebral Subject*.

Mirko D. Grmek, *Pathological Realities: Essays on Disease, Experiments, and History*. Edited, translated, and with an Introduction by Pierre-Olivier Méthot, Foreword by Hans-Jörg Rheinberger.

Lightning Source UK Ltd.
Milton Keynes UK
UKHW010912190321
380623UK00007B/173